iPhone®

2024 Edition

by Guy Hart-Davis

iPhone® For Dummies®, 2024 Edition

Published by: **John Wiley & Sons, Inc.,** 111 River Street, Hoboken, NJ 07030-5774, www.wiley.com

For general information on our other products and services, please contact our Customer Care Department within the U.S. at 877-762-2974, outside the U.S. at 317-572-3993, or fax 317-572-4002. For technical support, please visit https://hub.wiley.com/community/support/dummies.

Wiley publishes in a variety of print and electronic formats and by print-on-demand. Some material included with standard print versions of this book may not be included in e-books or in print-on-demand. If this book refers to media such as a CD or DVD that is not included in the version you purchased, you may download this material at http://booksupport.wiley.com. For more information about Wiley products, visit www.wiley.com.

Library of Congress Control Number: 2023947549

ISBN 978-1-394-22164-6 (pbk); ISBN 978-1-394-22166-0 (ebk); ISBN 978-1-394-22165-3 (ebk)

SKY10057247_101023

Contents at a Glance

Table of Contents

Icons Used in This Book

Little round pictures (icons) appear in the left margin throughout this book. Consider these icons miniature road signs, telling you something extra about the topic at hand or hammering a point home.

Here's what the five icons used in this book look like and what they mean.

TIP

This text contains the juicy morsels, shortcuts, and recommendations to make the task at hand faster or easier.

REMEMBER

This icon emphasizes facts and moves you'll likely benefit from retaining. You might even jot down a note to yourself in the iPhone's Reminders app.

TECHNICAL STUFF

These geeky sections are few and far between. They contain material that's interesting and informative, so you might want to look through them. You can safely ignore them if you want, because they won't be on the test.

WARNING

You wouldn't intentionally run a stop sign, would you? In the same fashion, ignoring warnings may be hazardous to your iPhone and (by extension) your wallet. There, you now know how these warning icons work, for you have just received your very first warning!

NEW

This icon marks a feature that's new in iOS 17 or the latest and greatest iPhones — as of this writing, the iPhone 15 family.

Beyond the Book

For details about significant updates or changes that occur between editions of this book, go to `www.dummies.com`, search for *iPhone For Dummies*, and open the Download tab on this book's dedicated page.

Also, the cheat sheet for this book has tips for mastering multitouch; a list of things you can do during a phone call; info on managing contacts; where to find additional help if your iPhone is acting contrary, and more. To get to the cheat sheet, go to `www.dummies.com` and type *iPhone For Dummies cheat sheet* in the Search box.

Where to Go from Here

Where to turn to next? Why, straight to Chapter 1, of course (without passing Go or collecting $200).

However, if you'd prefer to go straight to a particular topic, turn back a few pages to the table of contents and browse for the topic you want. Or if you want to look up something, head to the index at the end of the book to find which page you need to consult.

1

Up and Running with Your iPhone

IN THIS PART . . .

Set up and activate your iPhone, and then learn to navigate its user interface.

Master essential moves for making your iPhone do your will.

Load media from your computer and sync iCloud data across your iPhone, iPad, and computer.

Change key settings quickly with Control Center and control Siri with your voice.

Explore the Settings app and take full control of your iPhone.

Find the apps you need, install them, and manage them.

IN THIS CHAPTER

» Meeting your iPhone's hardware

» Setting up and activating your iPhone

» Starting to use your iPhone

» Locking and unlocking your iPhone

» Powering down your iPhone

Chapter 1

Setting Up and Navigating Your iPhone

Congratulations on getting an iPhone! You've made a great choice.

In addition to being a first-rate cellular telephone, the iPhone is the best iPod ever built, a gorgeous widescreen video player, and a fantastic camera and camcorder system, not to mention a powerful internet communications device.

This chapter starts by making sure you know your way around your iPhone's hardware. It then shows you how to activate the iPhone and set it up either manually or by picking up settings from your current iPhone or your iPad. You then learn to navigate the iPhone's Home screen pages and dock, lock the iPhone when you're not using it, and power it down for those rare occasions you don't need to keep it running.

Meeting Your iPhone's Hardware

On the outside, the iPhone's hardware is sleek and simple. This section explains what you find on the front, the back, the sides, and the bottom.

On the front

On the front of your iPhone, you find the following (labeled in Figure 1-1):

- **Camera:** The camera on the front of the iPhone is tuned for selfies and FaceTime video calling, so it has just the right field of view and focal length to focus on your face at arm's length, which presents you in the best possible light.
- **Receiver/front microphone:** The iPhone uses the receiver (speaker) and front mic for telephone calls. The receiver naturally sits close to your ear when you hold your iPhone in the "talking on the phone" position; the mic is used for noise cancelling and FaceTime calls.

  TIP

  If you require privacy during phone calls, use a compatible Apple or third-party headset — wired or wireless — as discussed in Chapter 5.
- **Status bar:** The status bar displays important information, as you discover later in this chapter.
- **Touchscreen:** The touchscreen dominates the front of the iPhone, enabling you to control the iPhone by gesturing with your fingers and thumbs.
- **Home button and Touch ID sensor (Touch ID models):** No matter what you're doing, you can press the Home button at any time to display the Home screen, which is the screen shown in Figure 1-1. The iPhone's Touch ID sensor uses your fingerprint to unlock the phone and to authenticate you (see Chapter 2).
- **App icons:** Each icon on the Home screen launches an app, such as the Weather app or the Files app.
- **Widgets:** A widget is a sort of mini-app that displays information or enables you to access features. For example, the Music widget in Figure 1-1 shows you current and upcoming songs.

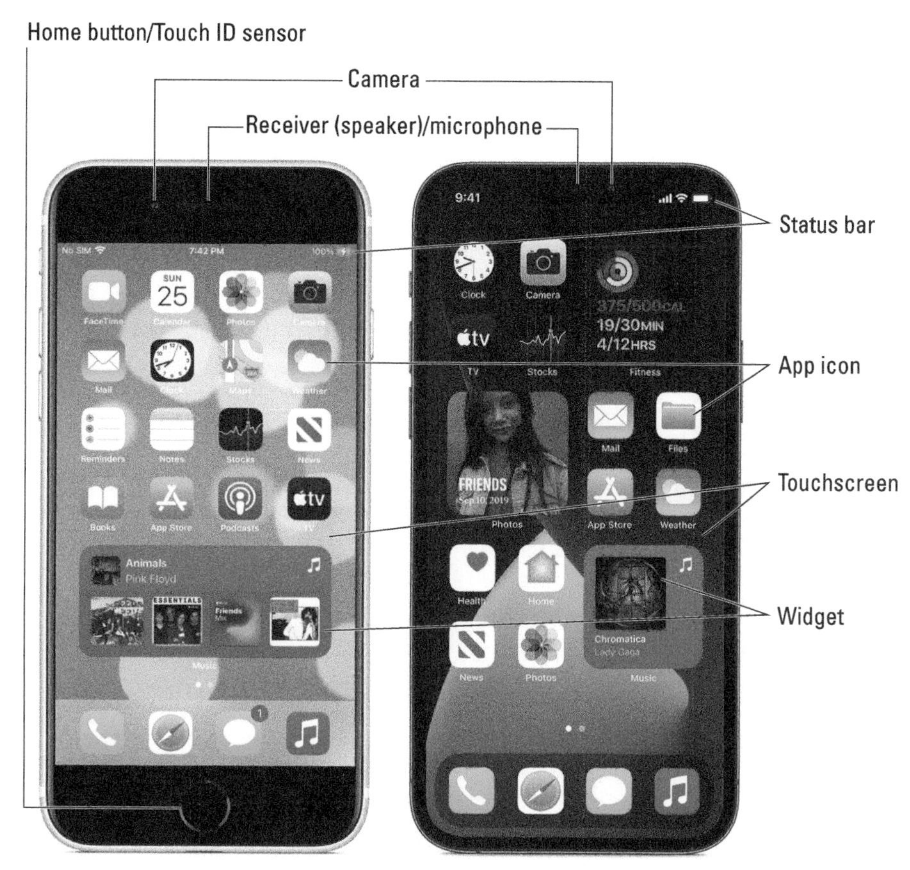

FIGURE 1-1: Touch ID iPhone models (left) have a Home button, whereas Face ID iPhone models do not.

Photo courtesy of Apple, Inc.

On the back

On the back of your iPhone are one to three camera lenses that look like little circles or ovals in the top-left corner. The iPhone also has one or more little LEDs next to the camera lens for use as a flash for still photos, as a floodlight for videos, and as a flashlight that you can turn on or off via Control Center. For more on using the camera and shooting videos, see Chapters 14 and 15, respectively; for more on the flashlight and Control Center, see Chapter 4.

On the sides

Here's what you'll find on the sides on your iPhone (see Figure 1-2):

- **Side button:** This physical button on the right side of the iPhone enables you to take several actions, including powering on your iPhone; putting it to sleep and waking it; and summoning Siri, the virtual assistant.
- **SIM card tray:** If your iPhone model uses a physical SIM card, open this tray, insert the card in it, and then replace the tray. iPhone 14 and iPhone 15 models sold in the US use an e-SIM (a virtual SIM) rather than a physical card. The SIM card tray may be on either the right side or the left side, depending on the iPhone model.

  TECHNICAL STUFF

  A SIM (Subscriber Identity Module) card is a removable smart card used to identify mobile phones. When you switch phones, you can move the SIM card from your old phone to the new phone, provided the phones use the same SIM card size. Current iPhone models that use a SIM card use the nano-SIM format.

  WARNING

  nano-SIM cards are tiny, so they're easily lost or damaged. Don't remove yours without good reason. If you do remove the SIM card, put it in a box or an envelope.

- **Ring/silent switch:** This switch on the left side of your iPhone lets you quickly toggle between ring mode and silent mode. When the switch is set to ring mode — the up position, with no orange showing on the switch — your iPhone plays all sounds through the speaker on the bottom. When the switch is set to silent mode — the down position, with orange visible on the switch — your iPhone makes no sound when you receive a call or when an alert pops up on the screen.

  NEW

- **Action button:** In place of the ring/silent switch, the iPhone 15 Pro models have the action button, which you can customize in the Settings app to take your preferred action, such as activating the Camera app or turning on the flashlight. The action button's default action is to toggle between ring mode and silent mode; long-press the action button until you feel haptic feedback confirming the mode change.

  REMEMBER

  Silent mode is overridden by alarms you set in the iPhone's Clock app; by music, audiobooks, and other audio you play; and by you auditioning sounds such as ringtones and alert sounds in the Settings app. Also, when you configure a focus, such as Do Not Disturb, you can permit specific apps to interrupt it.

  TIP

  If your phone is set to ring mode and you want to silence it quickly when it starts ringing, press the side button or either of the volume buttons.

» **Volume up/down buttons:** Two volume buttons are just below the ring/silent switch or the action button. The upper button increases the volume; the lower one decreases it. You use the volume buttons to raise or lower the loudness of the ringer, alerts, sound effects, songs, and movies. During phone calls, the buttons adjust the voice loudness of the person you're speaking with, regardless of whether you're listening through the receiver, the speakerphone, or a headset.

Volume up/down buttons
Side button
Ring/silent switch
(action button 15 Pro models)

FIGURE 1-2: Here's what you'll find on the sides of your iPhone.

The iPhone 14 Pro models and all iPhone 15 models have a feature called Dynamic Island, a resizable display element that appears near the top of the screen to provide context-sensitive controls and information, such as playback controls for music, telephony controls for phone calls and FaceTime calls, or map directions for your current journey. Tap an icon in Dynamic Island to go straight to the app for the feature that icon represents.

The iPhone 14 Pro models and the iPhone 15 models also have a new feature called Always-On Display that displays key information, such as notifications and upcoming events, on the Lock screen. On iPhone models without Always-On Display, locking the phone turns the screen off completely.

On the bottom

On the bottom of your iPhone, you find microphones, the Lightning port or USB-C port, and stereo speakers, as shown in Figure 1-3:

» **Microphones:** The built-in microphones let callers hear your voice when you're not using a headset.

The iPhone sports three or more microphones — the main ones are on the bottom — which work together to suppress unwanted and distracting background sounds on phone calls using dual-mic noise suppression and beam-forming technology. Beam-forming technology may make you smile, but its main purpose is to change the directionality of an array of microphones — in other words, to make them listen in the right direction.

» **Lightning port or USB-C port:** The Lightning port or USB-C port has several purposes:

- *Recharge your iPhone's battery.* Connect one end of the included charge cable to the iPhone and the other end to a USB power adapter, a USB port on your computer, or a port on a powered USB hub.
- *Sync your iPhone.* Connect one end of the cable to the port on your iPhone and the other end to a USB port on your Mac or PC.
- *Connect your iPhone to other devices.* Connect a camera, a television, or an external drive easily. If you have an iPhone 15 model, you can make the connection with a USB-C cable. If you have an earlier iPhone model, you will need a Lightning adapter such as the camera connection kit or one of Apple's A/V adapter cables.
- *Connect EarPods or a headset.* If you have Apple EarPods with a USB-C connector, you can plug them straight into the USB-C port on an iPhone 15 or later; similarly, you can connect Apple EarPods with a Lightning connector directly to the Lightning port on a Lightning-port iPhone. To use another headset, you may need to get Apple's USB-C-to-3.5mm Headphone Jack Adapter, Lightning-to-3.5mm Headphone Jack Adapter, or a functional equivalent.

» **Stereo speakers:** The speakers are used by the iPhone's built-in speakerphone and for playing audio — music or video soundtracks — when no headset is connected. They also play the ringtone you hear when you receive a call. All current iPhones have stereo speakers.

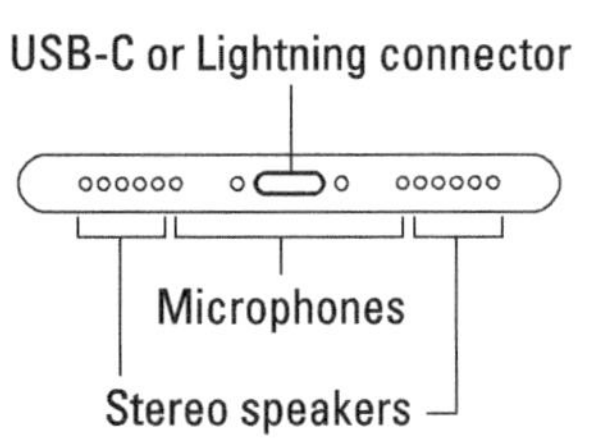

FIGURE 1-3: On the bottom of your iPhone (some models differ slightly).

Setting Up and Activating Your iPhone

Before you can start using your iPhone, you need to set it up, configuring its operating system (iOS) — the software that makes the iPhone tick — to work the way you want it to. You may also need to activate the iPhone to connect it to Apple's services and to your cellular carrier's service.

If you've already set up and activated your iPhone, skip this section.

Turning on your iPhone

Start by turning on your iPhone. Press and hold the side button, the physical button on the iPhone's right side. When the Apple logo appears on the screen, release the side button and wait until the Hello screen appears. This screen cycles through greetings in an apparently endless range of languages and their native scripts.

On the Hello screen, swipe up from the bottom of the screen on a Face ID iPhone or press the Home button on a Touch ID phone. The setup process begins.

Choosing the language, country or region, and appearance

A series of screens walks you through the first three steps of the setup process:

- **Language:** Choose the language you want your iPhone to use, such as English.
- **Country or Region:** Specify your country or region, such as United States.
- **Appearance:** Drag the slider along the Default-Medium-Large axis to choose the size of text and icons on the screen. Tap the Continue button.

At this point, the Quick Start screen appears, enabling you to finish setup quickly by using your current iPhone or iPad. Read on.

Using Quick Start or continuing setup

If you have an iPhone or iPad set up with your Apple ID, you can use that device to set up your new iPhone quickly. When the Quick Start screen appears (see Figure 1-4), unlock your current device and bring it close to your new iPhone. Your new iPhone then displays a complex pattern on the screen, and the Camera app opens on your other device, showing a target ring. Aim this ring at the pattern to establish the connection

FIGURE 1-4: From the Quick Start screen, you can swiftly set up your new iPhone using your current iPhone or iPad.

between the devices, and then authenticate yourself on your new iPhone by entering the current device's passcode. Follow the prompts to set up the new iPhone based on the current device. Skip ahead to the section "Starting to Use Your iPhone," later in this chapter.

If you have a current iPhone or iPad to use, or if you want to set up your new iPhone differently, tap the Set Up Without Another Device button, and then follow through the next subsection.

Setting up your iPhone without another device

If you chose to set up your iPhone without another device, work your way through the following screens:

- **Choose a Wi-Fi Network:** Tap the Wi-Fi network you want to use, type the password on the Enter Password screen, and then tap the Join button.

  If you need to use a Wi-Fi network that hides its network name, tap the Choose Another Network button. On the screen that appears, type the network name and then password, and then tap the Join button.

  If no Wi-Fi network is available, tap the Continue Without Wi-Fi button to use a cellular data connection.
- **Activation:** Wait while iOS activates your iPhone over the Wi-Fi connection (or the cellular connection, if you tapped Continue Without Wi-Fi). Activation may take several minutes.
- **Data & Privacy:** Read the information, and then tap the Continue button.
- **Face ID/Touch ID:** On a Face ID iPhone, follow the prompts to scan your face by aiming the recognition circle at your head, and then rotating your head to replace the white marks with green marks (see Figure 1-5).

FIGURE 1-5: Rotate your head to set up Face ID.

On a Touch ID iPhone, follow the prompts to scan the print of the finger or thumb you want to use to unlock the iPhone. You can add up to four more fingers (or thumbs — max two, preferably) after you finish setup.

» **Create an iPhone Passcode:** Tap the six-digit passcode you want to use, and then confirm it on the second screen.

If a six-digit passcode doesn't suit you, tap the Passcode Options button on the Create an iPhone Passcode screen. On the iPhone SE, you may need to scroll down to see the Passcode Options button. In the dialog that opens, tap Custom Alphanumeric Code, Custom Numeric Code, or 4-Digit Numeric Code, as appropriate, and then enter the code twice on the following screens. A custom alphanumeric code of eight characters or more is the most secure, though it will take longer to type. A custom numeric code enables you to create a numeric code of the length you prefer. Four digits is the minimum. Each digit more than six digits makes the code that much more secure. A four-digit numeric code is not strong enough for serious security, but you might want to use such a short code on an iPhone you're using for a demonstration, such as at a trade show.

» **Transfer Your Apps & Data:** Choose whether (and if so, how) to transfer your apps and data to your new iPhone. You have five choices:

- *From iCloud Backup:* Tap this button if you have a backup of your current or previous iPhone and want to restore it to your new iPhone.
- *From Another iPhone:* Tap this button if you have a current iPhone and want to transfer its apps and data. Normally, you would use the Quick Start feature earlier in the setup process to transfer your data from your current phone, but this button provides an alternative means.
- *From Mac or PC:* Tap this button if you used your Mac or PC to back up your previous iPhone and you want to restore data from one of those backups.
- *From Android:* Tap this button if your current device is an Android phone or tablet. You won't be able to transfer apps available only on Android.
- *Don't Transfer Anything:* Tap this button to set the iPhone up from scratch.

For the first four choices, follow the prompts in the resulting screens. For example, when restoring from an iCloud backup, you need to sign in to iCloud using your Apple ID, and then select the backup to use.

» **Terms and Conditions:** Read as much of the Terms and Conditions as you wish, and then tap the Agree button if you want to proceed.

» **Make This Your New iPhone:** This screen appears after you set up the means of transferring apps and data using one of the methods mentioned previously. The screen displays buttons summarizing what can be transferred, such as Apps & Data, Settings, and Wallet. If one of these buttons has > at its right end, you can tap the button to reveal a list of details — for example, tapping the Wallet button reveals a list of the payment cards transferred with Wallet. Tap the button again to hide the details. Tap the Continue button to continue with this selection, or tap the Customize button if you want to choose what to transfer.

- **Update Your iPhone Automatically:** Tap the Continue button if you want iOS to automatically download and install updates. Keeping iOS updated is good from a security perspective, but it means you may occasionally find your iPhone updating when you want to use it. If you prefer to control when iOS installs updates, tap the Only Download Automatically button instead.
- **Location Services:** Tap the Enable Location Services button to enable Location Services immediately. Normally, you'll want to enable Location Services because apps such as Maps depend on them. You can tap the Disable Location Services button if you don't want to use Location Services or if you plan to enable it later.
- **Apple Pay:** Tap the Continue button (and then follow the prompts) if you want to set up Apple Pay now, adding one or more credit or debit cards. Tap the Set Up Later button if you prefer to set up Apple Pay later or not at all.
- **Siri:** Tap the Continue button to set up the voice-driven virtual assistant now, or tap the Set Up Later in Settings button to set up Siri later (or never). If you enable Siri, the Improve Siri & Dictation screen appears, prompting you to share your Siri audio recordings anonymously to help Apple improve Siri; tap the Share Audio Recordings button or the Not Now button, as appropriate.
- **Screen Time:** Tap the Continue button if you want to activate iOS's parental-control and self-control feature now. Screen Time can be highly effective for tracking iPhone usage, either your own or that of your family members. However, if you don't plan to use Screen Time, tap the Set Up Later in Settings button instead.
- **iPhone Analytics:** Tap the Share with Apple button or the Don't Share button, as appropriate, to choose whether to share analytics data anonymously to help Apple improve the iPhone and iOS.
- **App Analytics:** Tap the Share with App Developers button or the Don't Share button, as appropriate, to choose whether to share app analytics data with developers, again anonymously.

Starting to Use Your iPhone

Once you've completed the setup routine, the Lock screen appears. Unlock it using Face ID or Touch ID:

» **Face ID:** Hold the iPhone pointing at your face.

» **Touch ID:** Place your registered finger on the Home button.

Meeting the Home screen

After you unlock your phone, the Home screen appears. It's divided into pages, with the first page appearing first. If you haven't customized the Home screen yet, the first page should look more or less like Figure 1-6.

FIGURE 1-6:
On the Home screen's first page, you see the status bar, the dock, widgets, and app icons.

These are the items on the first Home screen page:

» **Status bar:** This narrow horizontal strip appears across the top of each Home screen page; many apps also display it. The status bar displays icons that provide a variety of information about the current state of your iPhone. See the next section for details.

» **Wallpaper:** The wallpaper is the background in front of which the app icons and widgets appear.

» **Widgets and widget stacks:** A *widget* is a kind of mini-app that displays information on a single topic, such as the Weather widget showing the weather forecast and the Calendar widget showing the next calendar event in Figure 1-6. Widgets come in various sizes. You can arrange widgets of the same size into a *widget stack,* a virtual vertical stack in which you see only the topmost widget but can pull down on that widget to display the widget immediately below it.

- **App icons:** More than a dozen app icons appear by default on the first Home screen. These icons are for frequently used apps, such as Camera, FaceTime, Mail, Calendar, and Clock.
- **Search button:** The search button gives you instant access to the search feature.
- **Dock:** This area appears at the bottom of each Home screen page, giving you access to four apps no matter which page is displayed. The default apps are Phone (for phone calls), Safari (for web browsing), Messages (for instant messaging), and Music (three guesses).

Swipe left across the main section of the first Home screen page (not across the status bar or the dock) to display the second Home screen page (see Figure 1-7). You can see that the status bar, the search button, and the dock remain the same, but the selection of app icons changes. There is also a folder called Utilities, which contains several lesser but useful apps. To open the folder, tap it.

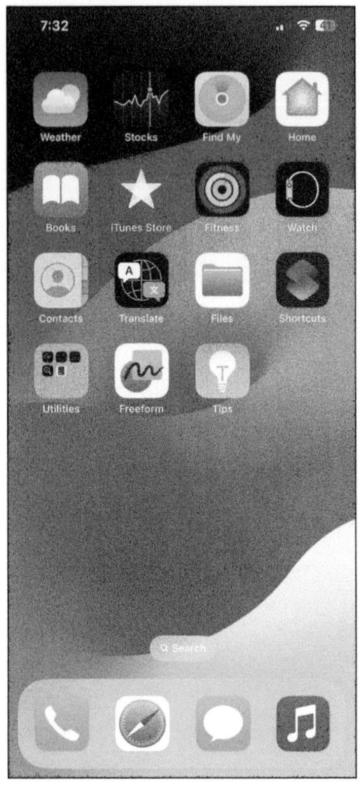

FIGURE 1-7: The second Home screen page.

Identifying the status bar icons

The status bar regularly displays icons showing the current time, the status and strength of the iPhone's cellular connection and Wi-Fi connection, and the proportion of battery power left. Beyond such widely useful information, the status bar can display a wide range of icons, depending on your iPhone's status. The following list shows you the status icons and explains what they indicate:

5GE

- **5GE:** Your wireless carrier's 5GE network is available.

4G

- **4G:** Your wireless carrier's high-speed UMTS network is available.

3G

- **3G:** Your wireless carrier's 3G UMTS or EV-DO data network is available and your iPhone can connect to the internet via 3G.

» **Airplane mode:** All wireless features of your iPhone — the cellular, 5G, 4G, 3G, GPRS (General Packet Radio Service), and EDGE networks, Wi-Fi, and Bluetooth — are turned off. However, you can turn on Wi-Fi and Bluetooth if you need them.

» **Alarm:** You've set one or more alarms in the Clock app.

» **Battery:** This battery icon displays the level of your battery's charge. The icon is completely filled with green or white when your battery is fully charged and then empties as your battery becomes depleted. If the icon is green, your iPhone is connected to a viable power source; if the icon is white or black, no power source is connected. A lightning bolt appears on the icon when your iPhone is recharging.

» **Bluetooth battery:** A tiny battery icon next to the Bluetooth icon displays the battery level of some Bluetooth devices.

» **Call forwarding:** Call forwarding is enabled on your iPhone, so it will forward incoming calls to the number you specified.

» **Camera in use indicator:** An app is using your iPhone camera. If you're using the Camera app, this won't be news, but the indicator can be helpful when other apps are using the camera.

» **CarPlay:** Your iPhone is connected to CarPlay, Apple's technology for making the iPhone work with the car's infotainment system.

» **Cell signal:** The strength of the cellular signal. The cell signal icon tells you whether you're within range of your wireless telephone carrier's cellular network and therefore can make and receive calls. The more bars you see (four is the highest), the stronger the cellular signal. If you're out of range, the bars are replaced with the words *No Service.* And if your iPhone is looking for a cellular signal, the bars are replaced with *Searching.*

If your screen shows only one or two bars, try moving around a little bit. Even walking a few feet can sometimes mean the difference between no service and three or four bars.

» **Do Not Disturb:** The Do Not Disturb feature (see Chapter 5) is enabled.

» **Driving:** The driving focus (see Chapter 5) is enabled.

» **Dual cell signal:** On iPhones with dual SIMs, the top row of bars indicates the signal strength of the line used for cellular data and the lower row of dots indicates the signal strength of your other line.

E

» **EDGE:** Your wireless carrier's slower EDGE (Enhanced Datarate for GSM Evolution) network is available and you can use it to connect to the internet.

GPRS

» **GPRS/1xRTT:** Your wireless carrier's slower GPRS data network is available and your iPhone can use it to connect to the internet.

- **Headphones connected:** Your iPhone is connected to Bluetooth headphones.

- **Location Services:** An application is using Location Services (see Chapter 12).

- **LTE:** Your wireless carrier's high-speed LTE network is available.

- **Microphone in use indicator:** An app is using your iPhone's microphone.

- **Network activity:** Some network activity is occurring, such as over-the-air synchronization, sending or receiving email, or loading a web page. Some third-party apps use this icon to indicate network or other activity.

- **Personal:** The personal focus (see Chapter 5) is enabled.

- **Personal hotspot:** This iPhone is connected to the internet via the personal hotspot connection of another device.

- **Personal hotspot indicator:** This iPhone is providing a personal hotspot connection or screen mirroring to another device; or an app is actively using your location.

- **Portrait orientation lock:** The iPhone screen is locked in portrait orientation. Open Control Center and then tap the portrait orientation lock icon to lock your screen in portrait orientation.

- **Reading:** The reading focus (see Chapter 5) is enabled.

- **Recording indicator:** Your iPhone is either recording sound or recording your screen.

- **Syncing:** Your iPhone is syncing with Finder or iTunes.

- **Sleep:** The sleep focus (see Chapter 5) is enabled.

- **TTY:** Your iPhone is set up to work with a teletype (TTY) machine, which is used by those who are hearing or speech impaired. You need an optional Apple iPhone TTY adapter (suggested retail price $19) to connect your iPhone to a TTY machine.

- **VPN:** Your iPhone is currently connected to a virtual private network (VPN).

- **Wi-Fi:** Your iPhone is connected to the internet over a Wi-Fi network. The more arcs you see (up to three), the stronger the Wi-Fi signal. If your screen displays only one or two arcs of Wi-Fi strength, try moving around a bit. If you don't see the Wi-Fi icon in the status bar, internet access via Wi-Fi is not currently available.

TECHNICAL STUFF

Wireless (that is, cellular) carriers may offer one of five data networks. The fastest (in theory) are the fifth-generation networks; the next fastest is 4G (LTE and UMTS); the next fastest is 3G; and the slowest are EDGE and GPRS. Your iPhone looks for the fastest available network. If it can't find one, it looks for a slower network.

Wi-Fi networks, however, are usually even faster than cellular data networks. So iPhones connect to a Wi-Fi network if one is available, even when a 5G, 4G, 3G, GPRS, or EDGE network is also available.

If you don't see any of these icons — 5G, LTE, 4G, 3G, GPRS, EDGE, or Wi-Fi — you don't currently have internet access.

Wi-Fi

- **Wi-Fi call:** Your iPhone is making a call over Wi-Fi.

- **Work:** The work focus (see Chapter 5) is enabled.

App Library and Home screen widgets

Two features that make finding what you need on your iPhone faster and easier are App Library and Home screen widgets; you learn all about both in Chapter 2.

Locking and Unlocking Your iPhone

When you're not using your iPhone, lock it and put it to sleep by pressing the side button once. Locking the iPhone like this turns the screen off, saving battery power, and prevents anyone else from using the iPhone without unlocking it. Locking the iPhone before you put it in your pocket or purse also helps ensure that the touchscreen doesn't register any bumps or touches that might change or delete something or even place an unwanted phone call.

TIP

Your iPhone automatically locks itself after a short period of inactivity, such as 30 seconds or a minute. To configure the length of time; choose Settings ⇨ Display & Brightness, tap the Auto-Lock button; and then tap the length of time, such as 30 Seconds, 1 Minute, or 5 Minutes.

Here's how to wake the iPhone from its sleep and unlock it:

- **Face ID iPhones:** Press the side button. Point the iPhone at your face and swipe up from the bottom of the screen. Face ID recognizes your face and unlocks the iPhone.
- **Touch ID iPhones:** Press the Home button using the finger or thumb you registered with Touch ID. The iPhone wakes, and Touch ID recognizes your fingerprint and unlocks the iPhone.

If the Raise to Wake feature is enabled, you can wake the iPhone by picking it up. To enable Raise to Wake, choose Settings ➪ Display & Brightness, and then set the Raise to Wake switch on (green).

Powering Down Your iPhone

Normally, you don't need to power your iPhone down. Instead, you simply put it to sleep when you're not using it, and then wake it up when you need it again.

When you do need to power your iPhone down, do so in one of these ways:

- » From the Home screen, choose Settings ➪ General ➪ Shut Down. The Power Off screen appears. Slide the Slide to Power Off slider to the right.
- » Press the volume up button, press the volume down button, and then press and hold the side button for a couple of seconds. The Power Off screen appears. Slide the Slide to Power Off slider to the right.

IN THIS CHAPTER

- » Tapping, swiping, dragging, and other gestures
- » Typing with the onscreen keyboards
- » Multitasking and switching apps
- » Organizing your Home screen pages
- » Keeping alert through notifications and today view
- » Using Standby

Chapter 2
Mastering Essential Moves

In this chapter, you get up to speed working with the iPhone's bright and beautiful touchscreen display. You start by learning the key touchscreen gestures, from tapping, double-tapping, and triple-tapping through to swiping, sliding, and pinching or spreading. Next, you explore the onscreen keyboards, multitasking, and switching apps, before moving on to navigating the Home screen pages and organizing the icons and widgets they contain. After that, it's time to search your phone, use notifications, and make the most of today view. Finally, you learn to customize the Lock screen and use iOS 17's new Standby feature.

Tapping, Swiping, Dragging, and More

To control your iPhone, you gesture with your fingers and (sometimes) thumbs on the touchscreen. The following list explains the eight main gestures:

- **Tap:** You tap the screen, placing your finger on it briefly and then lifting your finger again, to select items or to give commands. For example, you tap an app's icon on one of the Home screen pages to open that app. Similarly, you tap to start playing a song in the Music app, and you tap to open a photo album in the Photos app.
- **Double-tap:** You tap the screen twice in rapid succession to take actions such as zooming in and out on web pages, maps, and email messages.
- **Triple-tap:** You tap the screen three times in rapid succession to give special commands, such as enabling the Zoom feature, which lets you zoom the whole screen rather than zoom in individual apps.
- **Flick:** You flick your finger across the screen to scroll quickly through lists of songs, emails, and picture thumbnails. To flick, place your finger on the screen and then move it rapidly in the direction you want the content to move. For example, flick up a list of songs to move the list up so that you can see later items. You can either wait for the list to stop scrolling or tap to stop the scrolling.
- **Pinch and spread:** On a web page or picture, pinch your fingers together to shrink the image, or spread your fingers apart to enlarge the image. Pinching and spreading (sometimes called *unpinching,* occasionally *paunching*) are easy and effective gestures.
- **Drag:** Place your finger on the touchscreen and then, without lifting your finger, move it. You might drag to move around a map that's too large for the iPhone's display area.
- **Swipe:** Swiping is like a more controlled version of flicking; you place your finger on the screen and move it quickly but not extravagantly. For example, you can swipe left on the first Home screen page to display the second Home screen page.
- **Slide:** Sliding is a move you use with the onscreen keyboard's QuickPath feature, which lets you enter a word by placing your finger on the first letter and then sliding your finger to each other letter in turn without lifting it from the screen. When you finish the word, or when the Predictive feature guesses it correctly, you lift your finger, and iOS enters the word.

TIP

If you find it hard to reach the top of the screen when you're using the iPhone with one hand, you can use the Reachability feature. From the Home screen, choose Settings ➪ Accessibility ➪ Touch, and then set the Reachability switch on (green). You can then swipe down on the bottom edge of a Face ID iPhone, or double-tap (not double-press) the Home button on a touch ID iPhone, to shift the top half of the screen output down to the lower half of the screen, making it easier to reach. Reachability is especially helpful with larger-screen iPhones but works with all iPhones.

Getting the Hang of the Onscreen Keyboards

When you activate an app or a field that can accept text input, your iPhone automatically displays the onscreen keyboard so that you can type text. The keyboard has multiple variations on the alphabetical keyboard, the numeric and punctuation keyboard, and the more punctuation and symbols keyboard. The top row of Figure 2-1 shows three keyboards in the Notes app, while the second row shows three keyboards in Safari.

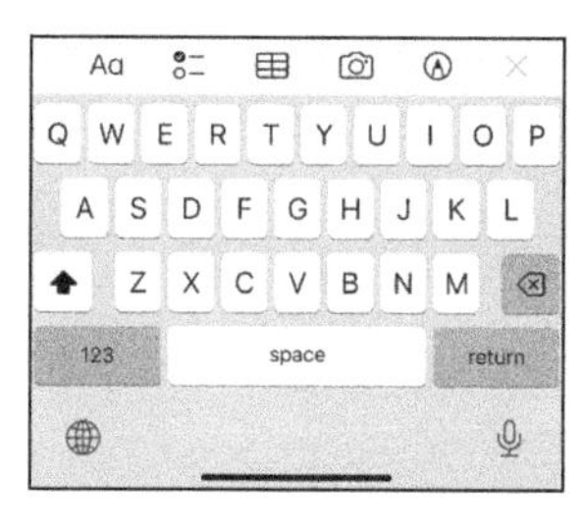

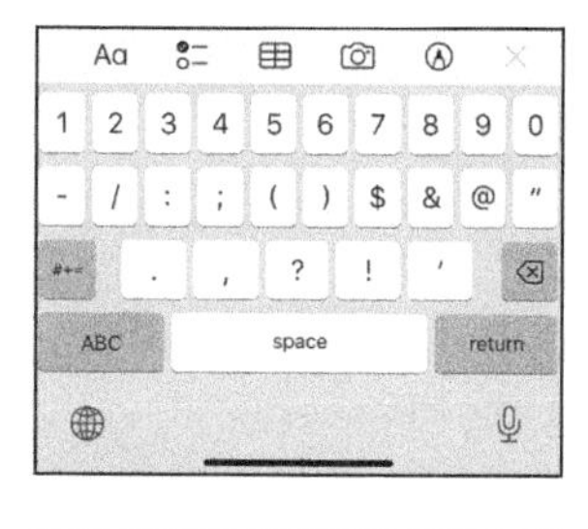

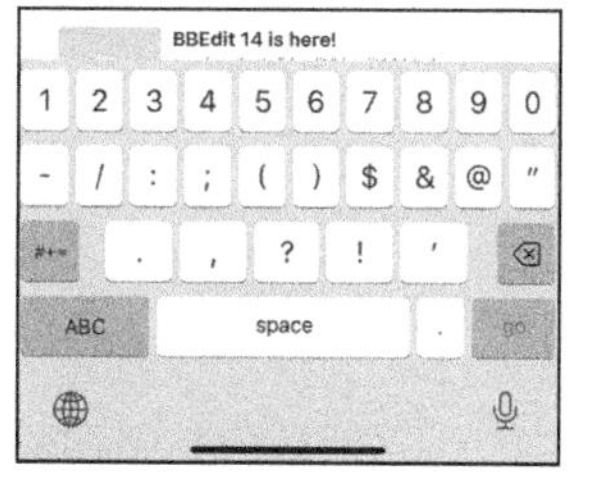

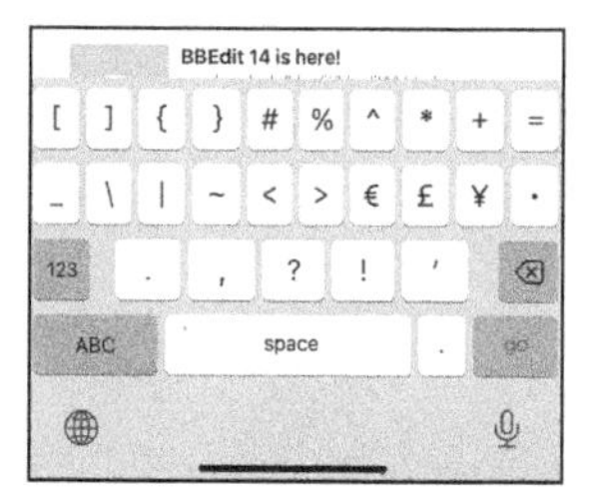

FIGURE 2-1: Six faces of the iPhone keyboard in portrait mode in the Notes app (top) and Safari (bottom).

The layout you see depends on which toggle key you tapped and the app that you're working in. If you look closely, you can see that the keyboards in Safari differ from the keyboards in Notes, sometimes in subtle ways. For example, in Figure 2-2, the Notes keyboards have a Return key in the lower right, but the Safari keyboards have a Go key in that position.

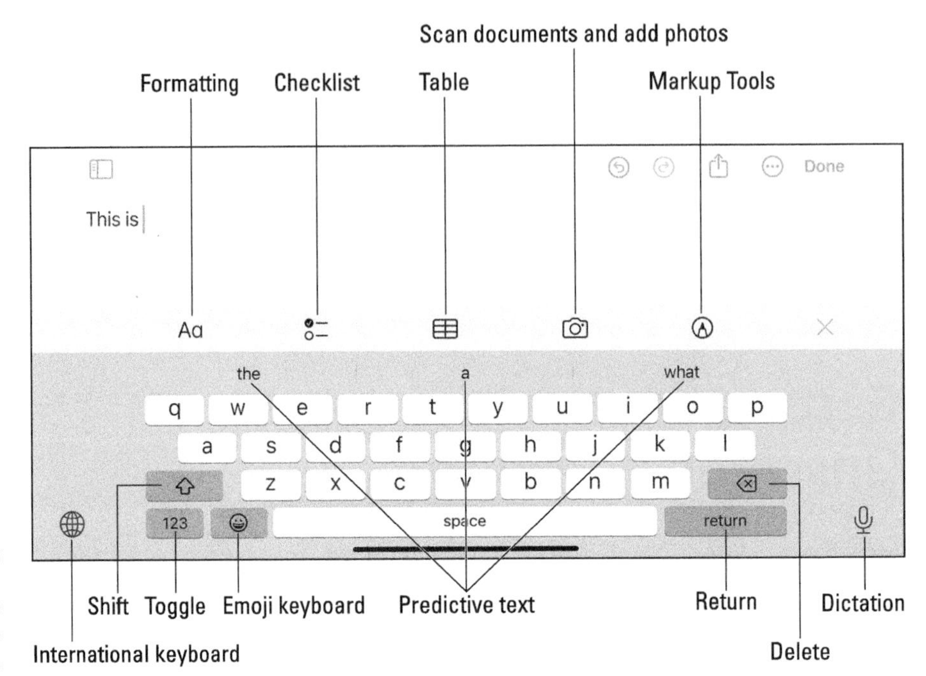

FIGURE 2-2: The keyboard in landscape mode in the Notes app.

When you rotate the iPhone to its side, you'll get wider variations of the respective keyboards. Figure 2-2 shows a wide version of an alphabetical keyboard in the Notes app.

Understanding the special-use keys

The iPhone keyboard contains a number of keys that don't type a character (refer to Figure 2-2). These special-use keys are

» **Shift key:** Tap to switch between uppercase and lowercase letters if you're using the alphabetical keyboard. If you're using keyboards that show only numbers and symbols, the traditional shift key is replaced by a key labeled #+= or 123. Tap that key to toggle between keyboards that have just symbols and numbers. When shift is on, the letters on the keys appear in uppercase.

TIP

When you need to type in all caps, double-tap the shift key to turn on caps lock. If this doesn't work, the Enable Caps Lock setting is turned off, and you need to turn it on. Choose Settings ➪ General ➪ Keyboard, and then set the Enable Caps Lock switch on (green).

» **Toggle key:** Tap to switch between the different keyboard layouts.

» **Emoji key:** Tap the smiley-face key to summon an emoji keyboard with, yes, smiley faces, among other emoticons. You'll find Apple's memoji stickers, custom memoji stickers you've created, and pictures of bells, boats, balloons, and animals. You also have access to the keyboard-switching menu if the international keyboard key (see the next entry) isn't displayed. If the emoji key doesn't appear, choose Settings ➪ General ➪ Keyboard ➪ Keyboards ➪ Add New Keyboard, and then tap Emoji to add the emoji keyboard.

On the iPhone SE models, the emoji key appears to the right of the 123/ABC key. When iOS displays the international keyboard key, it replaces the emoji key, and you access emojis by long-pressing the international keyboard key and then tapping Emoji. On other iPhone models, the emoji key appears below the keyboard. When iOS displays the international keyboard key, that key appears below the keyboard, and the emoji key appears to the right of the 123/ABC key.

- **International keyboard key:** This key, which bears a globe icon and is sometimes called the globe key, appears only when you've configured your iPhone to use multiple keyboards, as explained in the sidebar titled "A keyboard for all borders," later in this chapter. Long-press this key if you want to choose a one-handed keyboard, as discussed in the next section.

  Note: When you select a keyboard in a different language — or English for that matter — you can select different software keyboard layouts (such as QWERTY, AZERTY, and QWERTX). You can also specify a different layout for a hardware keyboard you connect via Bluetooth or the iPhone's USB-C port or Lightning port.

- **Delete key:** Erases the character immediately to the left of the insertion point.

TIP

If you hold down the delete key for a few seconds, it begins erasing entire words rather than individual characters.

- **Return key:** Moves the insertion point to the beginning of the next line. As mentioned, the Return key becomes a Go key on the Safari keyboard.
- **Dictation key:** Lets you use Siri to dictate your words. More on Siri in Chapter 4.

The iPhone can automatically recognize the language in which you're dictating based on the international keyboards you set up on your device.

Above the keyboard is the predictive text bar, which shows predictions for the next word. You can tap a prediction to enter it. When you're typing a word, you can press the spacebar to enter the middle prediction.

Above the predictive text bar, some apps display a bar of controls. For example, in Figure 2-2 you can see the Notes app's controls for applying formatting, creating a checklist, creating a table, scanning a document or adding a photo, and marking up an image.

One-handed keyboard

Apple makes typing easier for one-handed typists by adding a crunched keyboard that you can push to either the right or left side of the screen. To display this keyboard, long-press the globe key (labeled *International keyboard* in Figure 2-2) or the emoji key, if the globe key isn't visible. On the pop-up panel, tap the icon for the one-handed keyboard you want.

A KEYBOARD FOR ALL BORDERS

iOS includes international keyboard layouts for more than 120 languages. To add a keyboard layout, tap Settings ➪ General ➪ Keyboard ➪ Keyboards ➪ Add New Keyboard, and then tap the keyboard you want.

Have a multilingual household? You can add as many international keyboards as you want. When you're working in an app that displays the onscreen keyboard, choose the keyboard you want in one of these ways:

- Tap the international keyboard key to cycle through the keyboards.
- Long-press the international keyboard key to display a pop-up list of keyboards, and then tap the one you want.

You can use handwriting character recognition for simplified and traditional variations of Chinese. Just drag your finger in the box provided. Some Chinese keyboards don't rely on handwriting.

When you've added keyboards for multiple languages, you may be able to type in the two languages you use most frequently without having to switch keyboards. iOS supports using pairs of languages for languages such as English, French, German, Italian, Portuguese, Spanish, and Chinese. Just type the word or words in whichever language makes sense, and the iPhone guesses what you mean to type next — in the appropriate language.

To set your Preferred Language Order list, choose Settings ➪ General ➪ Language & Region. In the Preferred Languages box on the Language & Region screen, grab a language by the handle (three horizontal lines) on its right side, and then drag it up or down.

Anticipating what comes next

Whether you type by tapping or by using the Quick Path sliding method explained earlier, iOS tries to predict the next word, displaying its predictions on the predictive text bar above the keyboard and putting the most likely prediction in the middle. Say you're creating a note in the Notes app. You start typing, *This is*. The predictive text bar shows three predictions: *the*, *a*, and *what* (see Figure 2-2).

To enter one of these predictions, tap it. To enter the middle prediction when you've started typing a word, press the spacebar. Otherwise, keep typing, and the three predictions change to feature the letters you've typed so far.

Such keyboard predictions vary by app and by whom you are communicating with. So the predictions that show up in Messages when you're involved in an exchange with a friend are likely to be more casual than those in an email to your boss.

If you're using an international keyboard, predictions appear in the language you're using.

FIGURE 2-3: The Character Preview feature expands the key you tapped.

TIP

If you're typing by tapping rather than sliding, start with the index finger of your dominant hand, and then try graduating to using two thumbs. If you're using a one-handed keyboard, you'd normally type with your thumb.

When you tap a character on the keyboard, the individual key you press gets bigger, as shown in Figure 2-3, so you can see whether you tapped the right character. If you don't see the key get bigger, choose Settings ➪ General ➪ Keyboard and set the Character Preview switch on (green).

TIP

Sending a message to an overseas pal? Keep your finger pressed against certain letters, and a row of keys showing variations of the character for foreign alphabets pops up, as shown with the letter *e* in Figure 2-4. Slide your finger until you reach the key with the relevant accent mark, and then release it.

FIGURE 2-4: Typing an accented letter.

TIP

Long-press the period key in Safari or in a Mail address field to display a pop-up panel containing top-level domains, such as .com, .net, .edu, .org, and .us. Long-press the period key in regular text fields to get the choice of a period or an ellipsis.

If you enabled any international keyboards, you'll see other choices when you hold down the period key. For example, if you enabled a French keyboard, pressing and holding down the period in Safari gives you options including .eu and .fr.

When you make a typing mistake, such as typing *nptes* instead of *notes* (see Figure 2-5), iOS suggests *notes* as the most likely word, putting it in the middle of the predictive text bar, where you can accept it by typing a space. To keep what you typed, tap the left prediction. To use the prediction on the right, tap it.

FIGURE 2-5: The predictive text bar recommends *notes* but also lets you keep the typo by tapping it.

To type a number, symbol, or punctuation mark, tap the 123 key. The keyboard displays numbers and symbols, with the #+= key replacing the 123 key. Tap the #+= key to display additional symbols. Tap the ABC key to return to the alphabetical keyboard; the 123 key then replaces the ABC key.

TIP

In many apps, you can display a landscape keyboard with larger keys by rotating the iPhone to landscape mode.

Correcting mistakes

The iOS Auto-Correction feature does a terrific job of fixing typos. So even when you realize you've mistyped, generally you'll get a better result from keeping typing (and letting Auto-Correction fix what problems it can) than stopping and correcting each typo. The Predictive feature helps further by guessing many words correctly when you're only partway through typing them.

FIGURE 2-6: To correct a queried word, tap it, and then tap the correct prediction.

You'll still need to make some corrections manually. When iOS thinks you've made a mistake, it may underline or highlight the suspect word (see Figure 2-6). Tap the queried word, and then tap the correct word either on the pop-up bar or on the predictive text bar. If there's no correct prediction, or no prediction at all, correct the word manually.

If the middle item on the prediction bar shows a different word than you want, keep typing the rest of the word. When you finish typing the word, if the middle item is still not the word you want, tap the left item on the prediction bar to enter the word you've typed.

Positioning the insertion point

When you're editing text, you'll need to position the insertion point in precisely the right place. You can position the insertion point in three ways:

- **Tap:** Tap a word to position the insertion point after that word.
- **Drag your finger:** Place your finger on the screen in approximately the right place. iOS places the insertion point there and displays a loupe showing an enlargement of the text and insertion point (see the left screen in Figure 2-7). Drag the insertion point to where you want it.
- **Use the keyboard area as a touchpad:** Long-press the spacebar until the keys go blank (see the right screen in Figure 2-7). Then slide your finger across the keyboard area to move the insertion point.

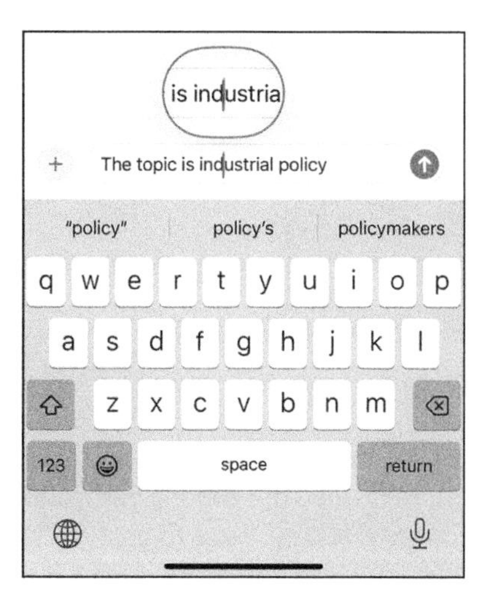

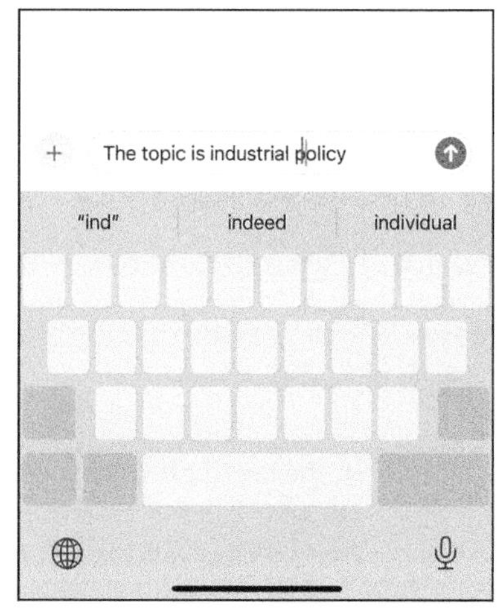

FIGURE 2-7: To position the insertion point precisely, either drag your finger on the screen and use the loupe for guidance (left) or use the keyboard area as a touchpad (right).

Choosing an alternative keyboard

Apple lets you install alternative keyboards from third-party app developers, such as the Microsoft SwiftKey keyboard and the Fleksy keyboard. Such keyboards may offer cosmetic differences (such as a different look or layout), enhanced features (such as predictions boosted by artificial intelligence, AI), or both.

To install another keyboard, open the App Store app and search either for a particular keyboard by name or for *iPhone keyboard* in general. Many keyboards are free to install, but watch out for the words *In-App Purchases*. They likely mean you'll need to pay either to unlock the app's most attractive features or to keep them beyond a trial period.

After you install another keyboard, choose Settings ➪ General ➪ Keyboard ➪ Keyboards, tap Add New Keyboard, and then select the keyboard. Back in an app that needs the keyboard, long-press the international keyboard key (the globe icon), and then tap the new keyboard on the pop-up list.

Cutting, copying, pasting, and replacing

iOS makes cut, copy, and paste easy. To copy (or cut) and paste, double-tap a word to select it, and then drag the start handle and the end handle to select more or less text, as needed (see the left screen in Figure 2-8). The pop-up bar appears automatically, giving you access to the Cut, Copy, Paste, and Replace commands. Tap Copy or Cut, as appropriate. Either way, the text goes on the Clipboard.

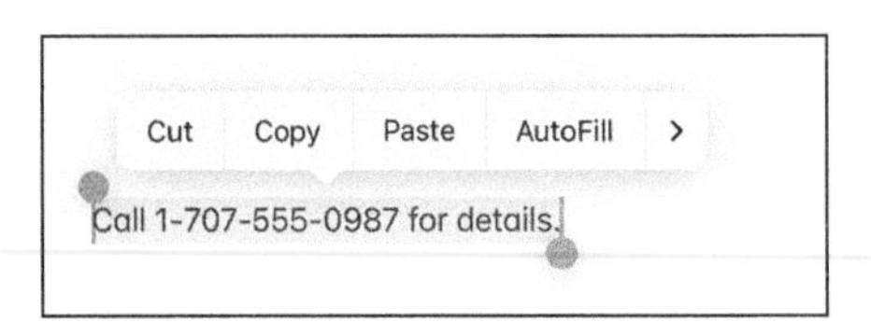

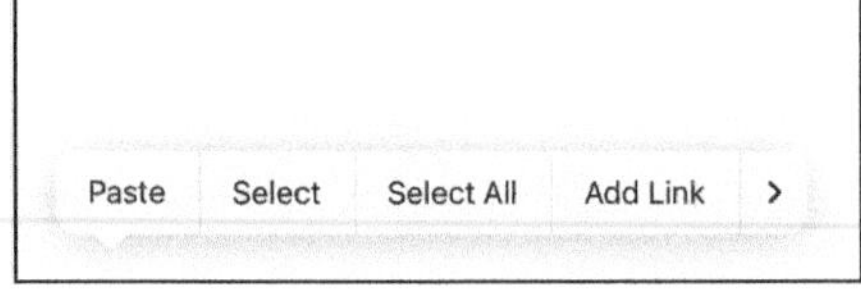

FIGURE 2-8: Drag the handles to select text, and then tap Copy (left). At the destination, tap Paste (right).

In the destination app, place the insertion point where you want to insert the copied text. Tap the insertion point to display the pop-up bar, which now includes the Paste, Select, and Select All commands. Tap Paste to paste in the text.

If you make a mistake, shake the iPhone to undo it. Alternatively, double-tap with three fingers.

If iOS underlines a word to indicate a possible typo, tap the word to see suggested replacements. Tap a correct prediction to enter it (see Figure 2-9).

FIGURE 2-9: Tap an underlined word to see suggested replacements.

To replace a word that's spelled correctly, double-tap it to display the control bar (see the left screen in Figure 2-10), and then tap > (more icon) to display the next batch of buttons (see the middle screen). Tap Replace to see suggested replacements (see the right screen), and then tap the word. (This is a tough choice, but I'm going with *monkey* rather than *Monet* or *honey*.)

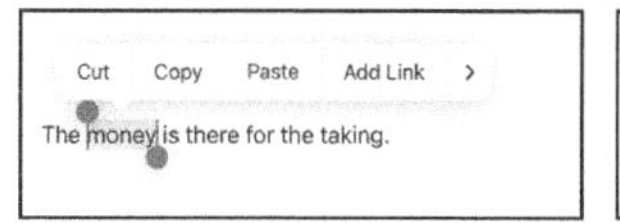

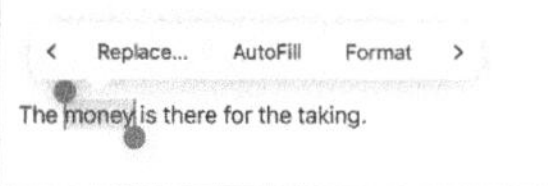

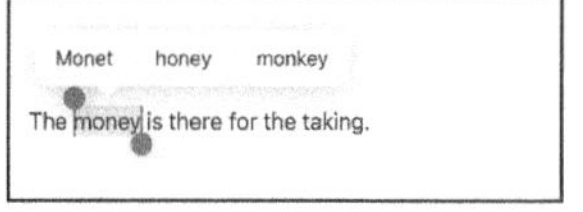

FIGURE 2-10: To replace a word that's spelled correctly, tap > (left), tap Replace (center), and then tap the word (right).

To find out exactly what a word means, double-tap the word, tap > until the Look Up button appears on the control bar, and then tap the Look Up command. The first time you tap Look Up, you're presented with the option to download the dictionary. Look Up also shows suggestions from iTunes and the App Store, along with movie showtimes and locations. It ties into the search capabilities of your iPhone.

iOS can also look up objects and text by using its Live Text feature. See Chapter 14.

Multitasking and Switching Apps

Multitasking lets you run numerous apps on your iPhone simultaneously and easily switch from one app to another. Normally, only one app is visible and is displayed full screen. This is the *foreground app.* All other apps are said to be in the *background,* where they keep running, but you don't see them. For example, the Music app can keep playing music in the background while you work in the Mail app in the foreground. You can switch quickly from one app to another, bringing a background app to the foreground and thereby moving the previous foreground app to the background.

The main exception to only one app being visible is that the picture-in-picture feature enables you to watch video or take part in a FaceTime call while working in other apps. The picture-in-picture video feed appears in a small window in front of the foreground app.

To switch from one app to another, you use App Switcher, which you display like this:

- » **Face ID iPhones:** Swipe up from the bottom of the screen, and then pause for a moment.
- » **Touch ID iPhones:** Double-click (not double-tap) the Home button.

App Switcher appears as a carousel containing previews of your open apps (see the left screen in Figure 2-11). The foreground app appears on the right, with the next most recently used app to its left, followed by other recently used apps in order. Each app's icon appears above its preview. Swipe from left to right to see more preview pages (see the right screen in Figure 2-11). Tap the icon or the preview for the app you want to switch to, and the app appears, enabling you to restart work or play where you left off.

To display the Home screen, tap it below the App Switcher's carousel.

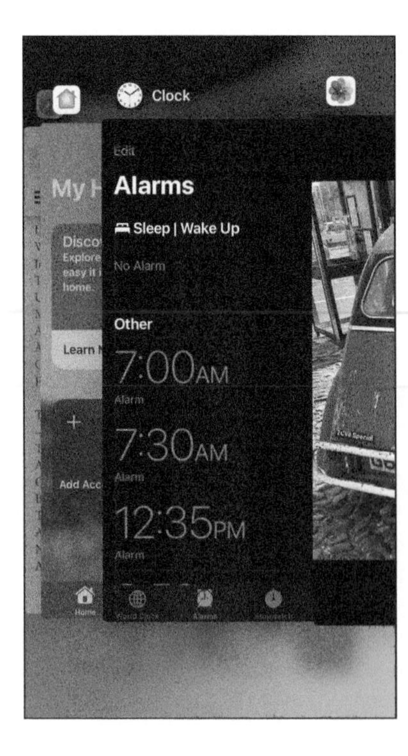

FIGURE 2-11: Scroll App Switcher to see previews of the apps you've recently used.

To close an app, swipe it up off the carousel. This move is especially useful when an app is not responding, but you can use it on any app at any time.

Navigating beyond the First Home Screen Page

The Home screen is divided into pages, one of which appears at a time. iOS normally starts you off with two Home screen pages, plus the App Library page, which appears after the last Home screen page. You can add other pages freely to organize your apps and widgets the way you prefer them. iOS also adds Home screen pages automatically when you install apps that overflow from the last existing page. You can have up to 15 Home screen pages.

The four icons in the bottom row — Phone, Safari, Messages, and Music by default — are in a part of the screen known as the *dock*. When you switch from one Home screen page to another as just described, these icons remain on the screen, unless today view, App Library, Control Center, or Notification Center is in view or you're inside an app.

By default, the oval search button appears above the dock, enabling you to search quickly from any Home screen page. This button does double-duty with a series of dots that indicate the number of Home screen pages (the total number of dots) and which page is currently displayed (the dot that is white rather than gray). The dots appear when you swipe left or right between Home screen pages.

TIP

If you want to see the dots all the time, choose Settings ⇨ Home Screen & App Library, go to the Search area, and then set the Show on Home Screen switch off (white). You can then display the Search panel by performing a short swipe down the middle of the Home screen.

If you swipe all the way from left to right, the Today screen appears; see the section "Using Today View," later in this chapter. Swiping all the way from right to left displays the App Library page; see the section "Visiting App Library," also later in this chapter.

You can easily move icons within a screen or from screen to screen. Long-press any icon until a menu appears. Tap Edit Home Screen on this menu — or simply continue to long-press — and all the icons on the screen will begin to jiggle. Then drag the icon you want to move to its new location. The other icons on the screen step aside to make room. To move an icon to a different Home screen page, drag it to the right or left edge of the screen and wait for the next page or previous page to appear. When you're satisfied with the new layout, tap the Done button on Face ID iPhones or press the Home button on Touch ID iPhones to stop the jiggling.

A circled minus sign also appears on each of the jiggling apps. Tap it if you want to remove the app from your phone. For a third-party app, you can remove it from the Home screen but leave it in App Library, or you can delete it. For either move, you'll get one last chance to change your mind. For most built-in apps, your only option is to remove them from the Home screen but leave them in App Library.

The menu that contains the Edit Home Screen command may also contain commands specific to that app. For example, Apple's Keynote presentation app's menu includes the New Presentation, Start with an Outline, and Go to Keynote Remote commands.

Want to jump back to the last Home screen page you used? Simply press the Home button on iPhones that have one, or swipe up on iPhones that don't. Want to jump to the first Home screen page, assuming you're not already there? Press Home or swipe up again.

On Face ID iPhones, press and hold down the side button for a second to invoke Siri. On Touch ID iPhones, press and hold down the Home button for a second.

Organizing Home Screen Icons into Folders

To organize the apps on your Home screen pages, you can create folders and add app icons to them. Like the Home screen itself, each folder can have up to 15 pages; each page can contain up to 9 icons; so a folder can contain up to 135 icons.

To create a folder, go to the Home screen page that contains the first two icons you want to put into a folder. (If they're on different Home screen pages, move one of them to the other's page.) Long-press one of those icons, and then tap Edit Home Screen on the pop-up menu, making all the icons on the screen jiggle. Drag the icon on top of the second icon, and iOS creates a folder for you, opening it and assigning it an automatic name based on the category of the two apps — for example, Productivity. To change the name, tap the *x*-in-a-circle to the left of the name, type a new name, and then tap Done on the keyboard.

Tap outside the folder to close it. You can then drag other app icons to the folder.

To launch an app that's inside a folder, tap that folder's icon, and then tap the icon for the app that you want to open.

You can drag apps into and out of any folder. If you drag all the apps outside the folder, or delete the last app in the folder, the folder automatically disappears.

Visiting App Library

App Library is a tool for storing and accessing apps you don't use so often. To find App Library, swipe from left to right on each Home screen page in turn.

At the top of the App Library screen is the search box, as you see in the left screen in Figure 2-12. Tapping in the search box makes App Library display its contents as an alphabetical list, as you see in the right screen in Figure 2-12.

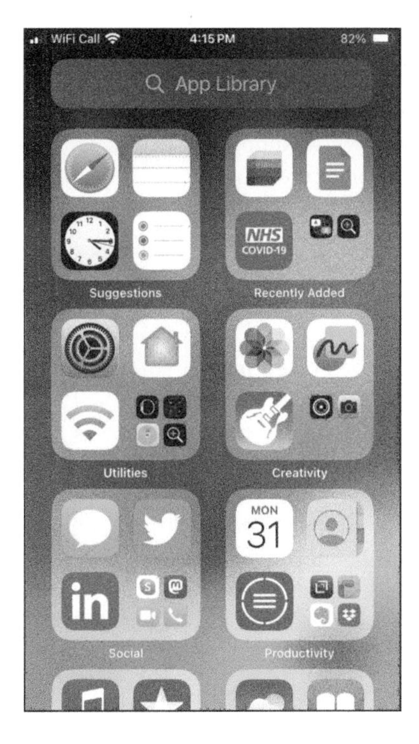

FIGURE 2-12: Checking out App Library.

You can scroll down to the app of choice, start typing the app name in the search box, or tap a letter on the side to jump to listings beginning with that letter. The # symbol (after Z in the list) takes you to apps whose names begin with a number.

Back on the initial App Library screen, just below the search box, iOS organizes apps into the Suggestions category and the Recently Added category. Suggestions contains suggested apps based on time of day, location, or activity. Below Suggestions and Recently Added, iOS presents the apps in various categories, such as Utilities, Creativity, Social, and Productivity.

Apart from Suggestions, each category contains three full-size icons and one group icon containing four miniature icons. Tap a full-size icon to launch that app. Tap the group of miniatures to display the remaining apps in the group. You can then launch an app by tapping its icon.

Working with Widgets

To enable you to glance quickly at discrete pieces of information, iOS lets you place widgets on the Home screen pages and in today view. For example, you could place a World Clock widget on a Home screen page so that you could see the time in four cities of your choice without opening the Clock app. Or you could place a Stocks widget to keep track of the Dow every minute of the trading day. Figure 2-13 shows these two widgets commandeering an entire Home screen page.

FIGURE 2-13: Widgets, such as the Clock widget and the Stocks widget, enable you to keep an eye on information directly from the Home screen or today view.

To add a widget, go to the Home screen page you want to place it on. Long-press anywhere on the page to make the apps jiggle, then tap + (add icon) in the upper-left corner of the screen. The Widgets overlay opens, and you can browse or search through the widgets. Drag a widget out of the overlay and drop it anywhere you want on a Home screen page or in today view.

From today view, you can start adding a widget by tapping Edit at the bottom of the screen.

You can also drag one widget on top of another to create a *widget stack*. You can then flip through these stacks by dragging your finger gently up or down. Inside a stack, iOS displays what it thinks is the right widget based on the time, location, or activity.

You can also create *smart stacks* that show up at the right time. For example, you might see widgets for the News app when you wake up, the Calendar app as an appointment time nears, and a Fitness app summary come evening. To create a smart stack, open the Widgets overlay, and then choose the Smart Stack item.

Searching Your iPhone

The Search button that appears on each Home screen page by default enables you to search your iPhone's contents quickly. Tap the Search button to display the Search screen (see the left screen in Figure 2-14). Look quickly at the Siri Suggestions box and the items below it to see if either shows what you want. If so, tap the item; if not, start typing your search term. Search results appear (see the right screen in Figure 2-14), and you can tap the search result you want to see.

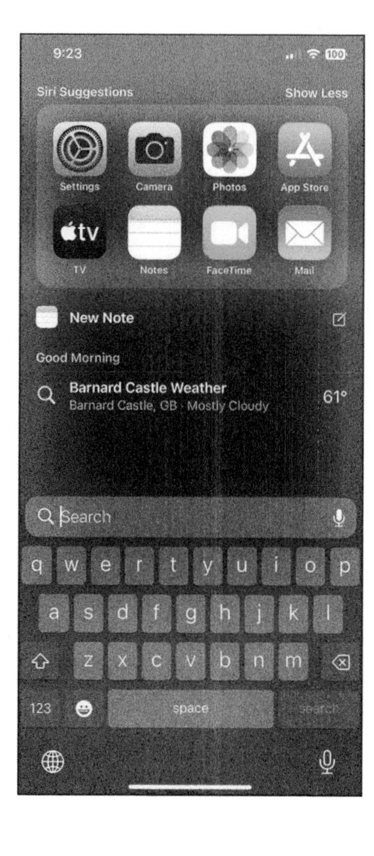

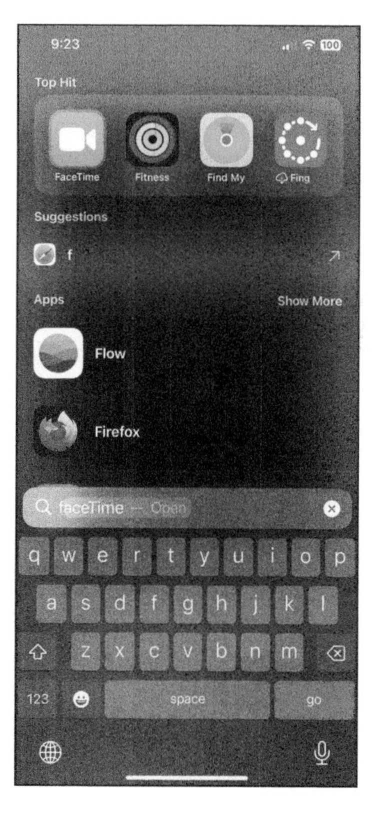

FIGURE 2-14: The Siri Suggestions box on the Search screen (left) shows apps you may want to run. If not, start typing your search term (right), and search results appear.

TIP

If you chose not to display the Search button on the Home screen, swipe a short distance down the middle of the Home screen to open the Search panel.

TIP

If your searches produce too many results, you can limit the search scope by choosing Settings ⇨ Siri & Search, and then working on the Siri & Search screen.

Using Notifications

To keep you in the loop, your iPhone displays notifications when particular events appear. When the iPhone is locked, you can display notifications in three ways on the Lock screen:

- **Stacked view:** Notifications appear in a stack near the bottom of the screen.
- **List view:** Notifications appear in a list, as shown in Figure 2-15. List view enables you to quickly triage your notifications for emergencies, but it may give other people a glance at your notifications too.
- **Count view:** The Lock screen displays the number of notifications waiting for you but discreetly doesn't show any specifics.

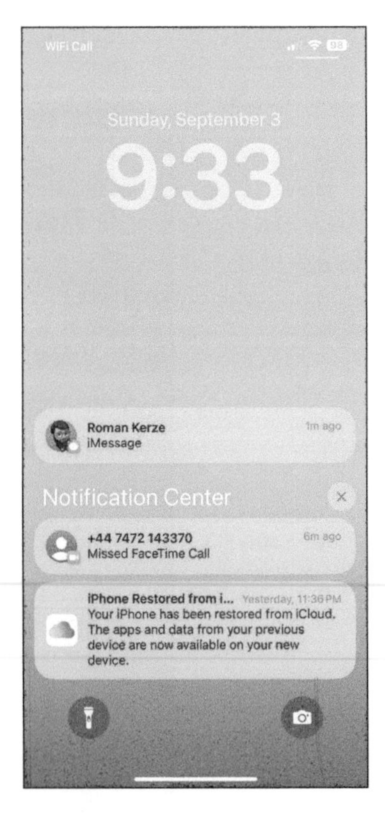

FIGURE 2-15: Staying in the loop with notifications.

When notifications are visible, swipe from right to left on a notification to clear or dismiss it. Swipe from left to right instead to open an underlying app, such as Mail, to enable yourself to deal with the cause of a notification immediately.

TIP

See Chapter 5 for instructions on configuring notifications.

You can also place Live Activities on the Lock screen. Live Activities include up-to-the minute sports scores, fitness workout stats, or notifications from your food delivery or Uber driver.

Using Today View

Today view, which you access by swiping left to right on the Lock screen or the first Home screen page, presents a collection of widgets designed to give you an overview of what's happening (or should be happening) today. Figure 2-16 shows the top part of today view with its default selection of widgets: The Photos, Calendar, Weather, and News widgets appear here, with other widgets further down the screen.

To get the most out of today view, customize it to contain only the widgets you want, and put them in your preferred order. To start customizing today view, scroll right down to the bottom of the screen and tap Edit. You can then remove an existing widget by tapping its remove icon (–), add a widget by tapping the add icon (+) in the upper-left corner of the screen, and drag your widgets into your preferred order.

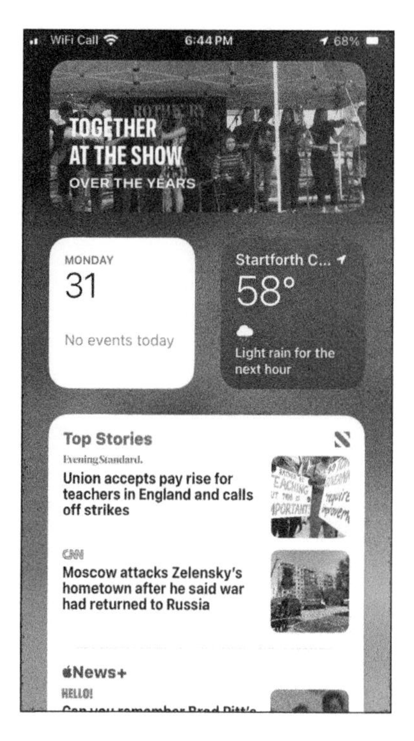

FIGURE 2-16:
Today view's widgets display information for the current day.

Customizing the Lock Screen

The Lock screen used to be merely a barrier to keep unauthorized people from using your iPhone. More recently, however, it has gained importance as a place to display widgets that allow you to keep tabs on key information without having to unlock your iPhone.

To customize the Lock screen, first display the Lock screen by locking your iPhone and waking it. Then long-press the Lock screen to switch it to Customization mode (see the left screen in Figure 2-17).

Tap + (add icon) to display the Add New Wallpaper screen (see the right screen in Figure 2-17), and then tap the wallpaper you want to use. On the resulting screen, swipe left to see different color treatments, and select the one you want. Next, tap the clock readout to display the Font & Color panel; choose the font, size, color, and intensity; and then tap x (close) to close the panel.

FIGURE 2-17: From the Lock screen (left), tap Add (+) to start creating a new custom Lock screen (right).

Next, tap Add Widgets to open the Add Widgets panel, and choose which widgets to place on the Lock screen. Bear in mind that these will be visible to anyone who can view or pick up your iPhone. Tap x (close) after making your choices.

When you've finished making your choices, tap + (add) in the upper-right corner to add the wallpaper to your collection. In the pop-up pane, tap Set as Wallpaper Pair to use the same screen for both the Lock screen and the Home screen, or tap Customize Home to create a custom version for the Home screen.

To switch among your custom Lock screens, long-press the Lock screen to switch to customization mode, and then swipe left or right. Tap the Lock screen you want to use.

Using Standby

New in iOS 17, the Standby feature enables you to use your iPhone as a desk clock and widget display while it's locked. Figure 2-18 shows the default Standby view, a large Clock widget and a Calendar widget showing a month.

FIGURE 2-18: Standby enables you to use your iPhone as a desk clock and widget display.

To use Standby, you need a charging stand that holds the iPhone vertically in landscape orientation (wider than tall). When you put the iPhone on such a stand, Standby may trigger automatically, or you may need to tap the iPhone's screen.

Once Standby starts, you can swipe up on either the Clock widget or the Calendar widget to reveal more options and start customizing Standby.

IN THIS CHAPTER

» Syncing iPhone using Finder or the iTunes app

» Syncing contacts and calendars

» Syncing everything else

Chapter 3
Getting in Sync

After you master essential iPhone moves (in Chapter 2), you'll likely want to get some or all of the following onto your iPhone: contacts, appointments, events, bookmarks, music, movies, TV shows, podcasts, and photos.

You can get items onto your iPhone either by syncing with your Mac or PC using Finder or iTunes, or by syncing across the internet using iCloud. Syncing your data either way keeps it up to date everywhere, so a change you make on your iPhone is available on your iPad and your Mac or PC, not to mention on your Apple Watch.

To sync with a Mac running a recent version of macOS (macOS 14, Sonoma; macOS 13, Ventura; macOS 12, Monterey; macOS 11, Big Sur; or macOS 10.15, Catalina), you use Finder; to sync with a Mac running macOS 10.14 (Mojave) or earlier or to sync with a PC, you use iTunes. If you're using iTunes, update to the latest version available by choosing iTunes ⇨ Check for Updates. Whether you're using Finder or iTunes, update iOS to the latest version too by choosing Settings ⇨ General ⇨ Software Update.

REMEMBER

The information in this chapter is based on macOS 14 (Sonoma) and iOS 17, the latest versions as of this writing, and uses Finder for most of the sync screens. If you're using macOS 10.14 (Mojave) or earlier or Windows, the screens you'll see in iTunes will look a bit different. But most things work in much the same way, so you'll have no problem following along.

You start this chapter by meeting iCloud, Apple's free (for your first 5GB, at least) online storage and synchronization solution, and find out how it makes using your iPhone more convenient. Then you get to the heart of this chapter: getting your stuff onto your iPhone.

If your iPhone is brand new and fresh out of the box, go to `www.dummies.com/how-to/content/setting-up-a-new-iphone.html` for a walk-through on the phone's initial setup.

Syncing via iCloud

The iCloud service is Apple's complete solution for syncing and storing your data online. iCloud can store and manage most (if not all) of your digital stuff — your music, photos, contacts, events, and documents— and make it available to all your computers and Apple devices automatically.

iCloud pushes information such as email settings, calendars, contacts, reminders, and bookmarks to and from your computer, iPhone, and iPad, and then keeps those items updated on all devices wirelessly without any effort on your part. iCloud is also the power behind the Files app (see Chapter 12).

Your free iCloud account includes 5GB of storage, an amount carefully calibrated to be enough to get you started, get you hooked on iCloud, and then get you paying to upgrade to iCloud+, the paid tier of iCloud, when you need more space. As of this writing, iCloud+ plans run $0.99 per month for 50GB, $2.99 per month for 200GB, $9.99 per month for 2TB, $30 per month for 6TB, and $60 per month for 12TB.

A nice touch is that all music, apps, periodicals, movies, and TV shows purchased from the iTunes Store, as well as your iTunes Match music (see Chapter 13) and your iCloud Photos content (see Chapter 14), don't count against your iCloud storage. Apple Books (e-books) don't count either, but audiobooks do.

Apart from audiobooks, what else counts against your storage? Most everything you can think of — mail, documents, photos and videos taken with your iPhone — and some things that probably don't spring to mind, such as account information, settings, and other app data. Some of these items take up little space, but photos get higher in resolution and larger in file size with each new iPhone model, and a 4K video can consume several gigabytes of storage in minutes. So if you want to stick with a free iCloud account, you'll need to police your storage vigorously.

To have your email, calendars, contacts, and bookmarks synchronized automatically and wirelessly between your computers and your Apple devices, enable iCloud syncing on your iPhone. Start by displaying the iCloud screen in the Settings app like this:

1. **On the Home screen, tap Settings to display the Settings screen.**
2. **Tap the button that bears your name at the top of the Settings screen.**

   This button is the Apple ID button. Tapping it displays the Apple ID screen.
3. **Tap iCloud to display the iCloud screen (shown in Figure 3-1, left).**

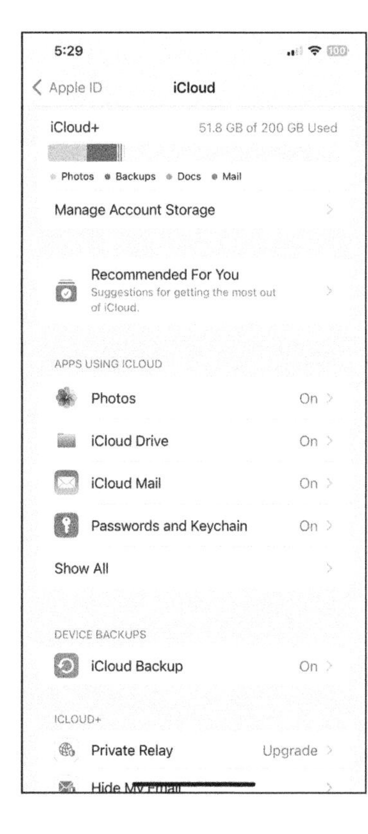

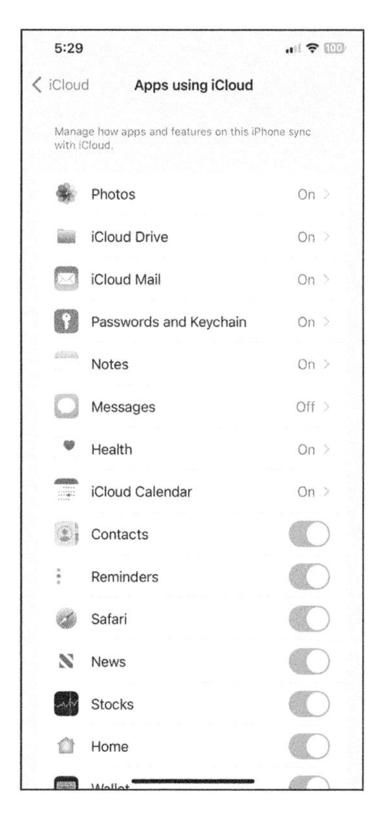

FIGURE 3-1: From the iCloud pane (left), you can check your usage of account storage and enable or disable an individual app's use of iCloud (right).

The iCloud screen contains five main items, all of which you'll benefit from knowing about:

- **iCloud histogram and Manage Account Storage button:** The histogram is a bar chart that shows how much of your iCloud space you're currently using. A readout gives you the text version, such as *23.6 GB of 50 GB Used*, and a multicolor bar breaks down the storage usage by category: Photos, Backups, Docs, Mail, and so on, all color-coded. Tapping the Manage Account Storage button takes you to the Manage Account Storage screen, where you can see how much space each app is using. You can even drill down into apps that have overstepped their boundaries and delete data to recover space.

- **Apps Using iCloud box:** This box is where you specify what to sync. You can tap any of the four items at the top — Photos, iCloud Drive, iCloud Mail, and Passwords and Keychain — to display a screen for configuring a sync for that item. Below these four items, tap the Show All button to display the Apps Using iCloud screen (shown in Figure 3-1, right), where the full list of apps appears. Some of these apps have a button you tap to display a configuration screen for the app, which includes an on/off switch for enabling or disabling sync; others have the on/off switch directly on the Apps Using iCloud screen.
- **Device Backups:** This box contains only the iCloud Backup button, which enables you to start backing up your iPhone with iCloud. See the next section.
- **iCloud+:** Enable the Private Relay feature to hide your iPhone's IP address and Safari browsing activity from all network providers and websites. This is a good move; so too is using the Hide My Email feature to hide your real email address and instead share a unique, randomly generated email address (such as `winsome_adroit.0x@icloud.com`), with any company or service you don't trust not to sell the address to spammers. The third feature, Custom Email Domain, enables you to use a custom domain name, such as `your_company.com`, with iCloud.
- **Manage Your Plan:** Tapping this button displays the Upgrade iCloud+ screen, where you can buy more iCloud storage. (If you've already bought more, you can downgrade.) You can also access this screen from the Manage Account Storage screen that you met at the top of this list.
- **Advanced Data Protection:** Tapping this button displays the Advanced Data Protection screen, where you can turn on advanced data protection. This feature applies end-to-end encryption to ensure that most iCloud data types, such as data from the Notes and Reminders apps, can be decrypted only by your trusted devices. This encryption provides strong protection, and you must set up a recovery contact or recovery key to make sure you can recover your data if you lose your trusted devices. Tap the Account Recovery button to set up a recovery contact or recovery key. Once you've done that, tap the Turn On Advanced Data Protection button to start the process of enabling the encryption.
- **Access iCloud Data on the Web:** Set this switch on (green) to enable access to your iCloud data via the `iCloud.com` website.

You find out more about iCloud in the rest of this chapter and several other chapters, so let's move on to syncing your iPhone by connecting it to your computer.

Syncing with Your Mac or Windows PC

You can sync your calendars, reminders, bookmarks, and other data and documents among your Apple devices and computers via iCloud, Finder (macOS Catalina and later), iTunes (macOS Mojave and earlier and Windows), or a combination. Connecting your iPhone to your computer not only lets you sync the two devices via Finder or iTunes but also enables you to create a local backup of your iPhone's contents.

If you don't have a computer, choosing iCloud as the means of backup is a no-brainer. But if you do have a computer, choosing between backing up to iCloud and backing up to your computer may be a tough decision. Bear in mind that restoring your iPhone from a computer backup requires a physical connection to the computer but doesn't require internet access, whereas restoring from iCloud does require internet access, and typically takes much longer — but you can restore your iPhone anywhere on Earth you can get online.

TIP

For the greatest safety and flexibility, back up both to iCloud and to your computer. You may want to create a third backup as well, one on a removable device in case your computer goes to the great bit-bucket in the sky.

What you need to know about iPhone backups

Your iPhone backs up your phone's settings, app data, and other information whenever you connect it to a computer and use Finder or iTunes to

- Sync with your iPhone
- Update your iPhone
- Restore your iPhone

Every time you sync your iPhone and computer, most (but not all) of your iPhone content — photos and videos you've taken with your iPhone (unless you're using iCloud Photos), text messages, notes, contact data, sound settings, and more — is backed up to wherever you choose — your computer's drive or iCloud — before the sync begins.

Most of your media, including apps, songs, TV shows, and movies, isn't backed up in this process because you can easily restore these files by syncing with your computer or redownloading them from the App or iTunes Store.

Backups are saved automatically and stored on your computer by default when you connect your iPhone to your computer with a USB-C cable (for USB-port iPhones) or a Lightning-to-USB cable (for Lightning-port iPhones).

If you don't want to involve your computer, you can choose to back up to iCloud, as you see in the next section.

Backups are handy if anything goes wonky with your iPhone or you get a new one. A backup lets you restore most of your settings and many files that aren't synced with iCloud or iTunes on your computer. If you've ever backed up an iPad or iPhone, you can restore the new iPhone with the older device's backup; the new device will inherit the settings and media from the old one.

TIP

When you set up a new device running iOS 17, you may see the Automatic Setup option, which lets you transfer settings from another iPhone or iPad to the new one. Both devices must be running iOS 11 or later and be within a couple of feet of each other during the setup process. Follow the on-screen instructions to transfer your settings (but not other content) to your new device.

Even if you take advantage of Automatic Setup, it's a good idea to set up your new iPhone to back up to either your computer or iCloud.

Here's how to enable backing up to iCloud from your iPhone. If you will use your iPhone without a computer, do this right away:

1. **On the Home screen, tap Settings to display the Settings screen.**
2. **Tap the Apple ID button, the button that bears your name at the top of the Settings screen, to display the Apple ID screen.**
3. **Tap iCloud to display the iCloud screen.**
4. **Tap the iCloud Backup button to display the Backup screen.**
5. **Set the Back Up This iPhone switch on (green).**

   Don't set the Back Up over Cellular switch on (green) unless you have an unlimited data plan.
6. **(Optional but advisable) To start a backup immediately, tap Back Up Now.**

WARNING

If you also intend to sync your iPhone with a computer, listen closely: Enabling iCloud Backup means your iPhone no longer backs up *automatically* when you connect it to your computer. You'll need to run backups to your computer manually by clicking the Back Up Now button in the Backups box on the General tab in Finder or the Summary tab in iTunes. This is easy to do; you just have to remember to do it.

TIP

Backups can consume a lot of disk space, especially if you have more than one iPhone or iPad. To check the amount of space left in your iCloud Storage, and to delete any backups you no longer need, choose Settings⇨Apple ID⇨iCloud on your iPhone, tap Manage Account Storage near the top of the iCloud screen, and look at the iCloud (or iCloud+) histogram at the top. The purple section shows how much space backups are occupying. To manage backups, tap the Backups button on the Manage Account Storage screen, and then work on the Backups screen. Here, you can tap a backup in the Backups list to display a screen that shows the backup's details and enables you to delete the backup.

If you're syncing your iPhone with your computer, click Manage Backups on the General tab in iTunes or the Summary tab in Finder to open a dialog in which you can see how much space your backups are taking up and delete backups you don't need.

Getting ready to sync

If you want to sync your iPhone with your computer, via either a cable or Wi-Fi, follow these steps:

1. **Connect your iPhone to your computer with the USB-C charge cable or the Lightning-to-USB cable.**

   Connect the iPhone to a USB port on the computer itself rather than a USB port on your keyboard, monitor, or hub. The computer's USB ports supply more power and have fewer points of failure.

   If you're using Finder, your iPhone appears in the Locations section of the sidebar. If you're using iTunes, the app launches.

   If that doesn't work, try the following, in order: Disconnect and reconnect the cable, plug in a different cable, restart your computer, and (macOS Mojave or earlier or Windows) launch or quit and relaunch iTunes.

2. **If your photo management software launches, either import the photos you've taken with the iPhone or don't.**

   If you've taken any photos with your iPhone since the last time you synced it, your default photo management software (Photos or Image Capture on the Mac; Photos on the PC) may open and ask whether you want to import the photos from your phone. (You find out all about this in the "Photos" section, later in this chapter.)

3. **If this is the first time you've connected your iPhone to your computer, do the following:**

   (a) *Answer the alerts.* An alert appears on each device, asking if you trust the other device.

   (b) *You do trust the devices, so click Continue on your computer, tap Trust on your iPhone, and enter your passcode if requested.* The Welcome to Your New iPhone pane appears.

   (c) *Choose the Restore from This Backup option or choose the Set Up as New iPhone option.*

   (d) *If you chose Restore from This Backup, select the most recent backup from the pop-up menu, and then click Continue.* Your new iPhone will be restored with the settings and data from your previous one. Soon, your new iPhone will contain all the apps, media files, and settings from your old iPhone. You're finished and can skip ahead to the next section.

   (e) *If you chose Set Up as New iPhone, click Continue, and then click Get Started.*

4. **Do one of the following:**

   - *macOS Catalina or later:* If you're not already seeing a Finder window displaying your iPhone's contents, open a Finder window manually. Click your iPhone's name in the Locations section of the sidebar, and then click General, as shown in Figure 3-2.
   - *macOS Mojave or earlier or Windows:* In the left sidebar of the iTunes window, click Summary, as shown in Figure 3-3.

     If what you're seeing looks different from Figure 3-3, make sure your iPhone is still connected. And if you see a drop-down menu that says Music, Movies, or anything else near the upper-left corner of the iTunes window, click the iPhone icon on its right (and shown in the margin) to begin working with your iPhone in iTunes.

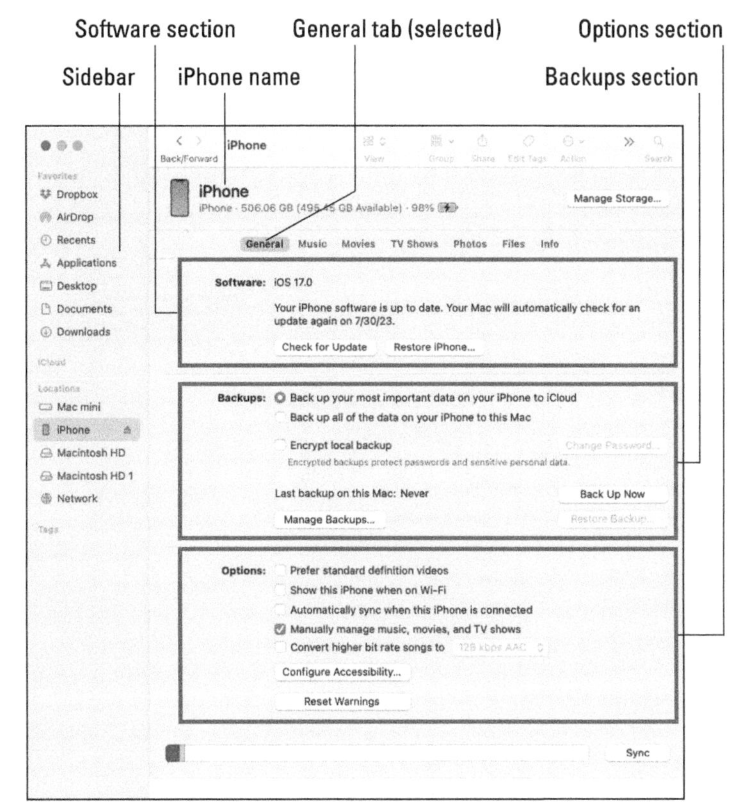

FIGURE 3-2: In the General pane in Finder, you can rename your iPhone, configure backup, and choose options.

If more than one iPhone or iPad is connected to this computer, click the iPhone icon to display a drop-down list and select your iPhone from the list.

If your screen still doesn't look like Figure 3-3 or you don't see an iPhone icon on the right of the drop-down menu and you're certain your iPhone is connected to a USB port on your computer (and not a port on a keyboard, monitor, or hub), try restarting your computer.

5. **If you want to rename your iPhone, click its name at the top of iTunes' sidebar on the left or at the top of the Finder window. Now select the old name and type a new one as shown.**

   The stunt iPhone in these figures is called simply iPhone.

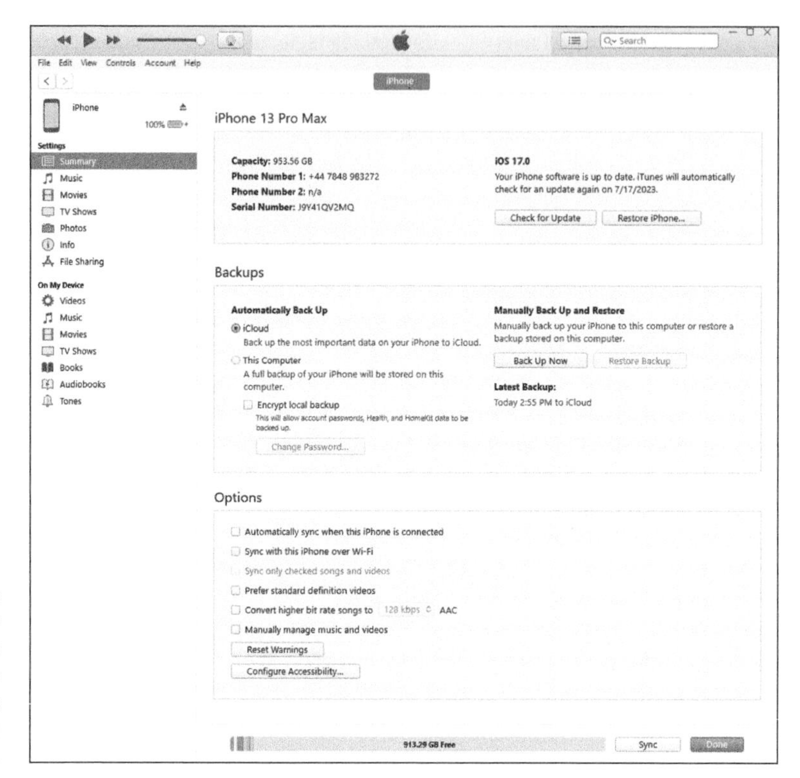

FIGURE 3-3: The Summary pane in iTunes contains similar controls to the General pane in Finder.

6. **In the Backups section of the General pane or Summary pane, click one of the following options:**

   - *Back Up Your Most Important Data on Your iPhone to iCloud:* Your iPhone creates a backup of its contents automatically every time you sync, regardless of whether you sync using a USB cable or wirelessly.
   - *Back Up All of the Data on Your iPhone (or PC) to This Mac:* If you choose this option, also select the Encrypt Local Backup check box. In the dialog that opens, type a strong password. On the Mac, select the Remember This Password in My Keychain check box to store the password safely in your Keychain. Then click Set Password.

TIP

Backups to iCloud are encrypted automatically, so if you chose iCloud, you're golden. But backups to your Mac or PC are not encrypted unless you select the Encrypt Local Backup check box. Here's the problem: If you don't encrypt the backup, your passwords, health data, and other sensitive information are *not* backed up. So normally you'll want to encrypt the backup.

Aside from the Back Up Now button and the Restore Backup button, that's all there is to the Backups section. The remaining steps deal with the check boxes in the Options section, which is below the Backups section (scroll down if needed).

7. **Do one of the following:**
   - *Finder:* To sync only items with check marks next to their names in your Music, Movies, and TV libraries, select the Manually Manage Music, Movies, and TV Shows check box.
   - *iTunes:* To sync only items that have check marks to the left of their names in your iTunes library, select the Sync Only Checked Songs and Videos check box (which is dimmed in Figure 3-3 because Apple Music is enabled).
8. **To sync your iPhone whenever you connect it to your computer, select the Automatically Sync When This iPhone Is Connected check box (in the Options section).**

WARNING

If you've selected the Prevent iPods, iPhones, and iPads from Syncing Automatically option in the Devices pane of iTunes Preferences (iTunes➪Preferences on a Mac; Edit➪Preferences on a PC), the Automatically Sync When This iPhone Is Connected option in the Summary tab appears dimmed and is not selectable.

TIP

If you select the Automatically Sync When This iPhone Is Connected check box, you can still prevent your iPhone from syncing automatically by holding down ⌘+Option (Mac) or Ctrl+Shift (PC) while you connect your iPhone. Keep holding down the keys until your iPhone's icon appears in the row of icons below the rewind/play/fast forward controls (iTunes) or the sidebar (Finder). This move prevents your iPhone from syncing automatically but doesn't change any settings.

9. **If you want high-definition videos that you sync to be automatically converted to smaller standard-definition video files when you transfer them to your iPhone, select the Prefer Standard Definition Videos check box.**

Standard-definition video files are significantly smaller than high-definition video files. You'll hardly notice the difference when you watch the video on your iPhone, and you can fit more video files on your iPhone because they'll take up less space. But if you watch video from your iPhone on an HDTV either with an A/V adapter cable or wirelessly via AirPlay (as discussed in Chapter 13), you'll notice a big difference.

If you have a wireless network at home, you don't need to sync video that you intend to watch at home with your iPhone. Instead, you can stream it from your computer to your iPhone or your iPhone to your HDTV (with an Apple TV).

WARNING

The conversion from HD to standard definition is time-consuming, so be prepared for long sync times when you sync new HD video with this option enabled.

10. **To sync automatically over your Wi-Fi connection, select the Show This iPhone When on Wi-Fi check box in Finder or the Sync with This iPhone over Wi-Fi check box in iTunes.**
11. **To turn off automatic syncing for music, movies, and TV shows, select the Manually Manage Music, Movies, and TV Shows check box.**
12. **To have iTunes automatically create smaller audio files (so you can fit more music on your iPhone), select the Convert Higher Bit Rate Songs To check box and choose a bit rate from the pop-up menu.**

TECHNICAL STUFF

A higher bit rate means that the song will have better sound quality but use more storage space. Most songs you buy at the iTunes Store or on Amazon, for example, have bit rates of around 256 Kbps. A 4-minute song with a 256 Kbps bit rate is around 8MB; convert the song to 128 Kbps AAC and it becomes roughly half that size (that is, around 4MB) while sounding almost as good.

Most people notice little (if any) difference in audio quality when listening to music on typical consumer audio gear. So unless you expect to hook your iPhone up to a great pre-amp and amplifier and superb speakers or headphones, you probably won't hear much difference with this option enabled, but your iPhone will hold roughly twice as many songs.

If you don't select the Automatically Sync When This iPhone Is Connected check box, you can synchronize manually by clicking the Sync button in the bottom-right corner of the window. If you've changed any sync settings since your last sync, the button is Apply instead of Sync.

Syncing Contacts and Calendars

Both Finder and iTunes enable you to sync your contact data and calendar data between your iPhone and your Mac or PC. This used to be the normal way of syncing data, but syncing via iCloud is now a better choice and is more widely used.

To see how your iPhone is configured for syncing contact data and calendar data, click the Info tab near the top of the Finder window or the Info category in the sidebar on the left side of the iTunes window, and then look at the Info pane that appears. Chances are you'll see messages saying that your iPhone is syncing contacts and calendars with iCloud. If so, you're set; don't make any changes.

If your iPhone is currently using iCloud to sync contacts and calendars, but you now want to switch to syncing using Finder or iTunes, you must turn off iCloud syncing for those services. On your iPhone, choose Settings ⇨ Apple ID ⇨ iCloud to display the iCloud screen, tap Show All to reach the Apps Using iCloud screen, and then set the switches for Contacts and Calendars off (white). You can then choose sync settings in the Info pane in Finder or iTunes.

Syncing Your Media

Next, let's look at how you get your media — your music, movies, TV shows, podcasts, video, and photos — from your computer to your iPhone.

Podcasts and videos are synced only from your computer to your iPhone, not vice versa. If you delete a podcast or a video you synced onto your iPhone, the next sync doesn't delete the podcast or video from your computer. But if you buy or download any of the following items on your iPhone, the item *will* be copied to your computer automatically when you sync:

- Songs
- Podcasts
- Videos
- E-books and audiobooks from Apple Books
- Playlists you've created on your iPhone

Syncing also copies to your computer any pictures you take with the iPhone's cameras, pictures you save from email messages or web pages (long-press the image, and then tap Save Image), and screen shots (see Chapter 17).

Music, movies, TV shows, and podcasts

Use the Music, Movies, TV Shows, and Podcasts panes to specify the media to copy from your computer to your iPhone. If you don't see these panes, make sure that your iPhone is still selected in the Locations list in the Finder sidebar (look back to Figure 3-2) or in the iTunes window (look back to Figure 3-3).

Music and music videos

To transfer music to your iPhone, select the Music pane, and then select the Sync Music check box. Figure 3-4 shows the Music pane in a Finder window in macOS Sonoma; if you're using iTunes, click Music in its sidebar, and you'll see the same options shown in Figure 3-4.

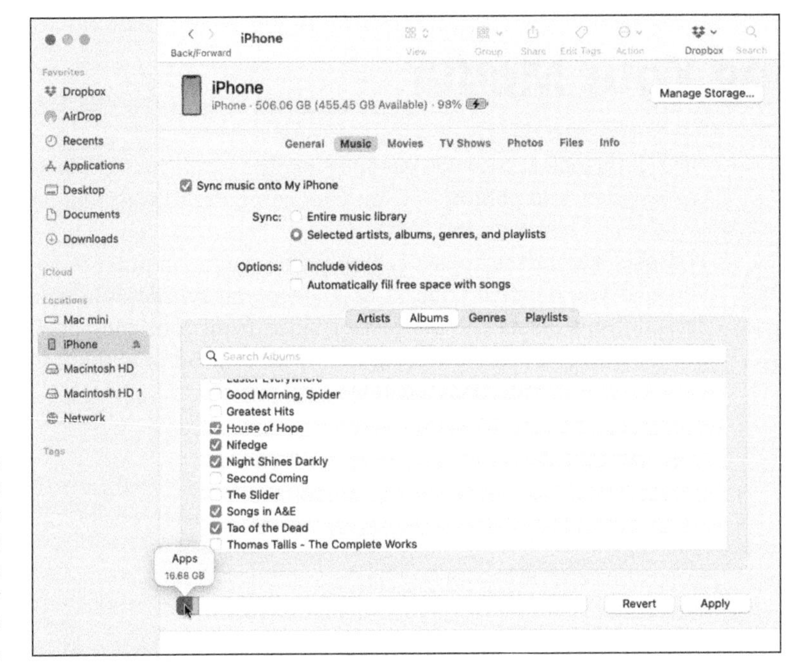

FIGURE 3-4: Use the Music pane to copy music and music videos from your computer to your iPhone.

You can then select the Entire Music Library option button or the Selected Artists, Albums, Genres, and Playlists option button, as shown in Figure 3-4. If you choose the latter, select the check boxes next to the particular artists, albums, genres, and playlists you want to transfer. Select the Include Videos check box at the top of the pane to include music videos.

WARNING

If you choose Entire Music Library and have more songs in your iTunes library than storage space on your iPhone, the sync will fail and the capacity bar at the bottom of the screen will display your overage. To avoid such errors, select playlists, artists, albums, and genres that total less than the free space on your iPhone, which is displayed in the capacity bar before you sync.

WARNING

Music, podcasts, and video use massive amounts of storage space on your iPhone. If you try to sync too much media, you'll see errors.

Hover your pointer over any color to see a bubble with info on that category, as shown in Figure 3-4.

You can find similar information about space used and space remaining on your iPhone by tapping Settings ➪ General ➪ iPhone Storage.

Finally, see the Automatically Fill Free Space with Songs check box? Don't select it. If you do, any free space on your iPhone will be filled with music. This was a good idea for the iPod but is a dreadful idea for your iPhone, which needs space free for storing photos and videos you take.

Movies

To transfer movies to your iPhone, click the Movies tab or pane, and then select the Sync Movies check box. You can also select the Automatically Include check box to choose an option for movies you want to include automatically, as shown in Figure 3-5 using Finder. If you choose an option other than All, you can optionally select individual movies and playlists by selecting the boxes in appropriate sections.

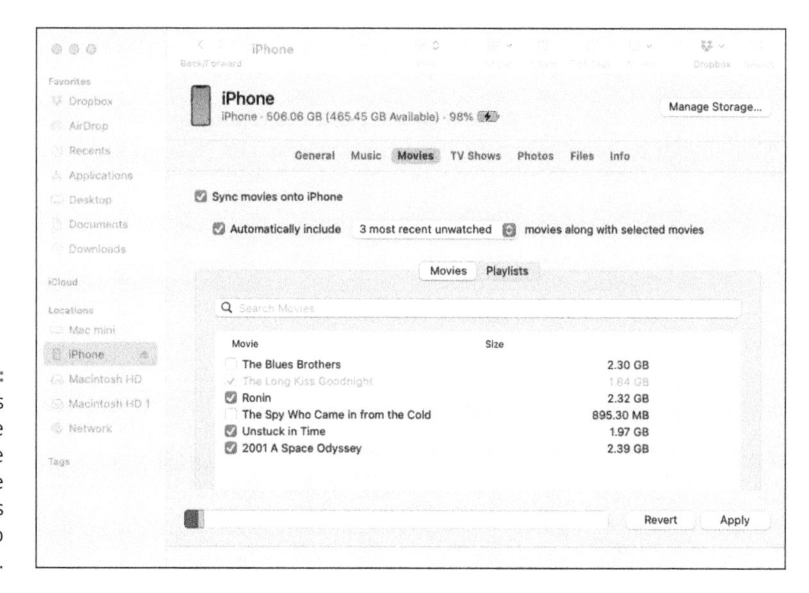

FIGURE 3-5: Your choices in the Movies pane determine which movies are copied to your iPhone.

TV shows, podcasts, books, and audiobooks

Syncing TV shows works like syncing movies. First, select the TV Shows pane or the TV Shows tab. Next, select the Sync TV Shows check box to enable TV show syncing. Then open the first pop-up menu and choose how many episodes to include. Next, open the second pop-up menu and choose either All Shows or Selected Shows. If you want to also include individual episodes or episodes on playlists, select the appropriate check boxes in the Shows, Episodes, and Include Episodes from Playlists sections of the TV Shows pane.

To transfer podcasts to your iPhone, select the Podcasts pane or the Podcasts tab and then select the Sync Podcasts check box. Next, you can include however many podcasts you want by making selections from the Automatically Include and All or Selected Podcasts pop-up menus.

You transfer e-books and audiobooks the same way — sync them all or select individual titles.

Photos

To sync photos between computers and Apple devices, you must enable iCloud Photos (formerly known as iCloud Photo Library).

You can also *copy* photos to your iPhone from the Photos app (Mac only) or any folder on your computer that contains images (Mac or PC).

To enable iCloud Photos:

- **On your iPhone:** Choose Settings ⇨ Apple ID ⇨ iCloud ⇨ Photos, and then set the Sync This iPhone switch on (green).
- **On your Mac:** In macOS Sonoma or Ventura, choose System Settings ⇨ Apple ID ⇨ iCloud, and then set the Photos switch on (blue). In macOS Monterey, Big Sur, or Catalina, choose System Preferences ⇨ Apple ID, and then select the iCloud Photos check box. In macOS Mojave or earlier, choose System Preferences ⇨ iCloud, and then select the iCloud Photos check box.
- **On your PC:** Download iCloud for Windows via the Microsoft Store app, and then launch it. Click Options (next to Photos), select iCloud Photos, click Done, and then click Apply. Now enable iCloud Photos on all your Apple devices.

  You can also choose to use My Photo Stream and iCloud Photo Sharing, and customize the location of your upload and download folders.

  When you turn on iCloud Photos on your PC, My Photo Stream is turned off automatically. If you want to send new photos to your devices that don't use iCloud Photos, you can turn My Photo Stream back on.

Now, connect the iPhone to your computer and return to the Photos pane or tab and select the Sync Photos check box. Next, choose an application or folder from the pop-up menu.

You can also type a word or phrase in the search field (an oval with a magnifying glass) to search for a specific event or events.

If you choose a folder full of images, you can create subfolders inside it that will appear as albums on your iPhone.

If you've taken any photos with your iPhone or saved images from a web page, an email, an MMS message, or an iMessage since the last time you synced, the appropriate program launches (or the appropriate folder is selected) when you connect your iPhone, and you have the option of uploading the pictures on your iPhone to your computer.

Syncing Items Manually

When you need to transfer just a few songs, movies, or whatever to your iPhone, use manual syncing. The process is as easy as dragging individual items to the appropriate location on your iPhone, as described next.

Automatic and manual sync aren't mutually exclusive. If you've set up automatic syncing, you can still sync individual items manually.

You can manually sync music, movies, TV shows, podcasts, books, and audiobooks but not photos and info such as contacts, calendars, and bookmarks.

To configure your iPhone for manual syncing:

1. **Connect your iPhone to your computer via USB.**

   If iTunes doesn't open automatically, open it manually (macOS Mojave or earlier or Windows).

2. **Select your iPhone:**

   - *Finder users: Click your iPhone in the sidebar.* If you have more than one Apple device, you'll see all connected iPods, iPhones, and iPads devices in the sidebar of all Finder windows.
   - *iTunes users: Click the iPhone icon to the right of the media kind drop-down menu.* If you have more than one Apple device, the iPhone icon becomes a drop-down menu listing all the devices. Click the icon to display the menu with your devices, and then select the device you want.

3. **To disable automatic syncing for music, videos, and TV shows, click the General or Summary tab, and then select the Manually Manage Music, Movies, and TV Shows check box in the Options section.**

TIP

If you're happy with automatic syncing and just want to get some audio or video from your computer to your iPhone, feel free to skip this step.

4. **Click the Apply button.**

Now, to actually sync individual items, do the following in Finder or iTunes: Click the appropriate tab or pane (Music, Movies, TV Shows, and so on) and enable the check boxes for the items you want to sync. If an item isn't available in the tab or pane you selected (ergo, no check box), locate its file on your computer's drive, drag it onto the appropriate tab, and then select its check box.

TIP

In iTunes, use the On My Device section in the sidebar to see which songs, movies, TV shows, and other media are already on your device. If the sidebar isn't displayed, choose View⇨Show Sidebar to display it.

REMEMBER

After lining up the files you want to sync, click the Sync (or Apply) button in the lower-right corner of the window to actually sync the files.

IN THIS CHAPTER

» Controlling your iPhone with Control Center

» Calling Siri

» Typing to request Siri

Chapter 4
Using Control Center and Siri

To get the most out of your iPhone, you'll want to use Control Center and Siri to the full. Control Center is a handy tool that gives you instant access to many of the most widely useful settings; Siri is the voice-activated personal assistant that Apple has built into the iPhone, the iPad, macOS, and even Apple Watch.

Controlling Control Center

Control Center is a quick-access panel for the controls, apps, and settings you frequently change. Control Center enables you to change settings swiftly and easily without having to visit the Settings app, but it also provides shortcuts to the relevant screens in the Settings app when you need to display them to take actions Control Center itself doesn't offer.

To access Control Center on iPhone models with Face ID, swipe down from the upper-right corner of the screen. On models with a Home button, swipe up from the bottom of the screen. Figure 4-1 shows Control Center. On a smaller-screen iPhone, you need to scroll down to see the controls at the bottom.

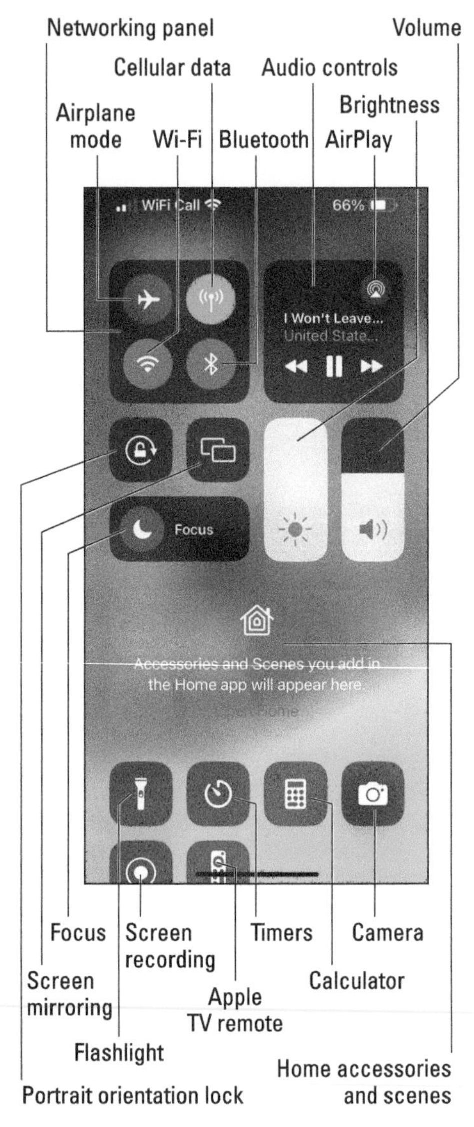

FIGURE 4-1: Control Center puts a stack of controls at your fingertips.

Here's what you need to know about the items in Control Center:

- **Networking panel:** This panel displays airplane mode, cellular data, Wi-Fi, and Bluetooth icons at first. Press the Networking panel to expand it, displaying two more icons — AirDrop and personal hotspot — plus the name of each icon and the feature's status, such as *Cellular Data On* or *Personal Hotspot Off*. Figure 4-2 shows the Networking panel expanded.
- **Airplane mode icon:** Tap this icon to toggle airplane mode on or off.
- **Cellular data icon:** Tap this icon to toggle cellular data on or off.
- **Wi-Fi icon:** Tap this icon to turn Wi-Fi off until tomorrow or to turn Wi-Fi back on. In the Networking panel, long-press the Wi-Fi icon to display the Wi-Fi overlay, which shows the list of Wi-Fi networks. Tap the network to which you want to connect, or tap Wi-Fi Settings to display the Wi-Fi screen in the Settings app, where you can take other actions.

» **Bluetooth icon:** Tap this icon to toggle Bluetooth off or on. In the Networking panel, long-press the Bluetooth icon to display the Bluetooth overlay, which shows a list of available Bluetooth devices. Tap the device to which you want to connect, or tap Bluetooth Settings to display the Bluetooth screen in the Settings app, which offers further actions.

» **AirDrop icon:** Tap this icon in the Networking panel to display the AirDrop dialog, where you can set your iPhone's AirDrop status by tapping Receiving Off, Contacts Only, or Everyone for 10 Minutes.

» **Personal Hotspot icon:** Tap this icon in the Networking panel to toggle Personal Hotspot between being discoverable by other devices or not discoverable. Even when personal hotspot is not discoverable, your other Apple devices signed into the same iCloud account can use the hotspot.

» **Portrait orientation lock icon:** Tap this icon to toggle the portrait orientation lock on or off.

» **Screen mirroring icon:** Tap this icon to display the Screen Mirroring list, which shows devices to which you can mirror your iPhone's screen.

» **Focus:** Tap the icon to toggle the default focus on or off. Tap the Focus label to display the list of focuses. Tap a focus to turn it on or off, or tap the ellipsis (...) to display different options, such as turning the focus on for one hour (see Figure 4-3). You can create a focus by tapping New Focus at the bottom of the screen.

» **Audio controls:** This panel shows the current or most recent audio source with the audio's name and minimal playback controls: previous/rewind, play/pause, and next/fast-forward. Tap the audio's name to go to the app, such as the Music app. Long-press the Audio controls box

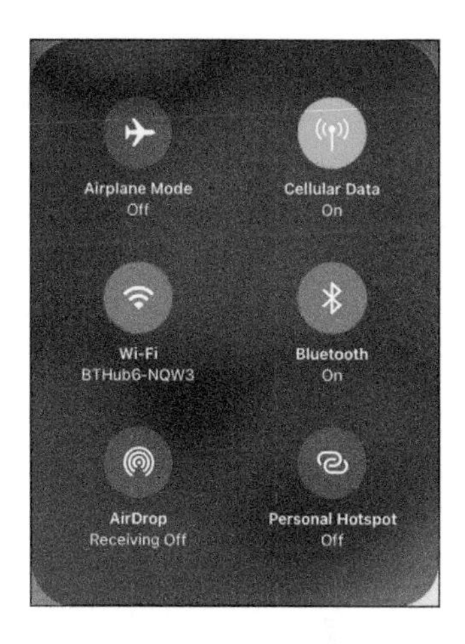

FIGURE 4-2: Expanding the Networking panel gives you access to AirDrop and Personal Hotspot.

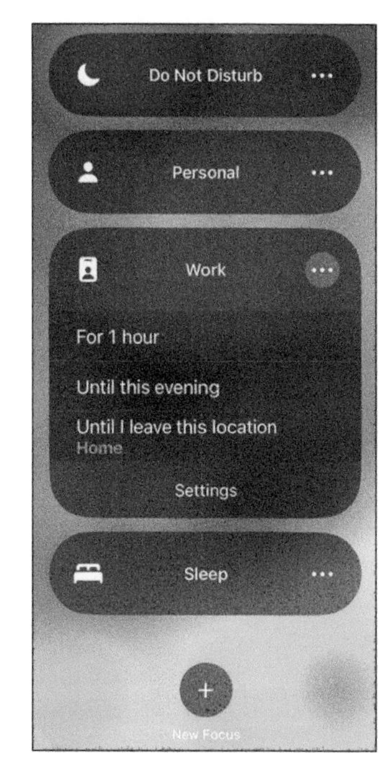

FIGURE 4-3: When you need peace, enable the appropriate focus.

to open the larger-size panel (see Figure 4-4). Here, you can drag the track position slider to move forward or backward through the track and drag the volume control to change the volume. For either control, you can put your finger anywhere on the bar and drag left or right — you don't need to drag the end of the filled-in section.

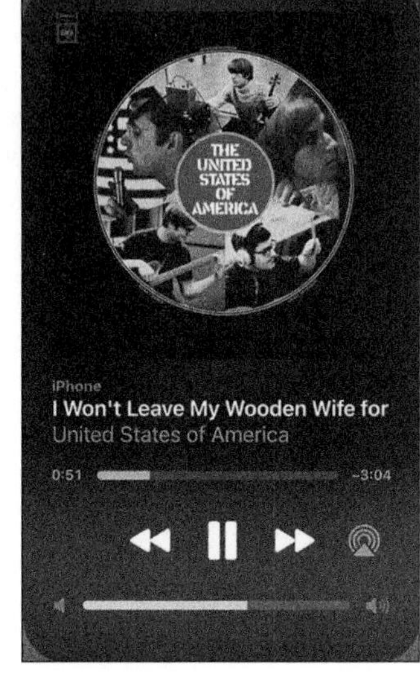

FIGURE 4-4: Expand the Audio controls panel to change the volume or move through the track.

- **AirPlay icon:** Tap this icon to display the audio controls and show the list of available AirPlay devices. Tap the device you want to use.
- **Brightness slider:** Drag this slider up or down to adjust the screen's brightness. Long-press the brightness slider to display the Brightness screen. This screen contains a larger version of the brightness slider, which can give you closer control, plus icons for toggling dark mode, night shift, and True Tone on and off.
- **Volume slider:** Drag this slider up or down to adjust the volume. Long-press the volume slider to display the Volume screen, which gives you a larger version of the slider for more precise adjustments.
- **Accessories and Scenes area:** Any accessories and scenes you add to the Home app appear here so you can access them quickly.
- **Flashlight icon:** Tap this icon to toggle the flashlight feature on or off. Long-press this icon to display the Flashlight screen, which enables you to change the flashlight's brightness.
- **Timers icon:** Tap this icon to display the Timers screen in the Clock app. Long-press this icon to display a control for setting and starting a quick timer.
- **Calculator icon:** Tap this icon to launch the Calculator app. If you just need to grab the result of the last calculation, long-press the Calculator icon, and then tap Copy Last Result on the pop-up panel.
- **Camera icon:** Tap this icon to launch the Camera app for regular use. To take a quick action, long-press this icon, and then tap the appropriate button in the Camera dialog: Take Selfie, Record Video, Take Portrait, or Take Portrait Selfie.
- **Screen recording icon:** Tap this icon to start a recording of what happens on screen. Screen recordings are useful for demonstrating techniques.
- **Apple TV remote:** Tap this icon to launch the Apple TV Remote app, which lets you control your Apple TVs from your iPhone.

Control Center's default set of controls can get you a long way, but you can also customize Control Center to better suit your needs. Choose Settings ➪ Control Center to display the Control Center screen in the Settings app (see Figure 4-5). Here, you can take five actions:

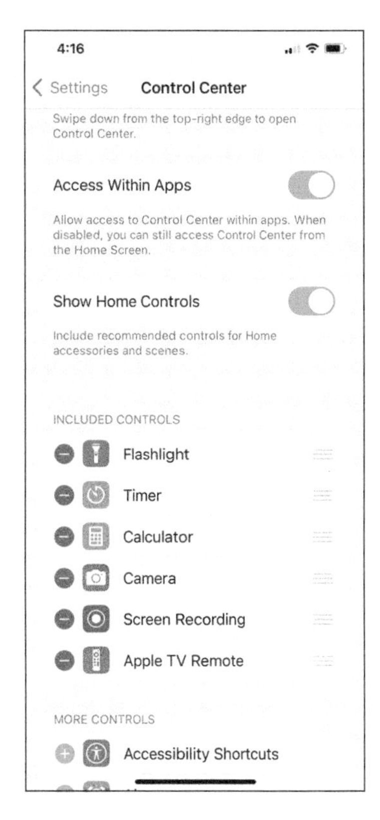

FIGURE 4-5: Use the Control Center screen in Settings to customize Control Center.

- **Enable or disable Access within Apps.** Set the Access within Apps switch on (green) or off (white), as needed. Usually, accessing Control Center within apps is helpful, but you might want to disable access if you find you invoke Control Center by mistake.
- **Show or hide the Home Controls.** You can suppress the Home controls by setting the Show Home Controls switch off (white) — useful if you don't use the Home app much.
- **Add controls.** In the More Controls list, tap the + in a green circle to move a control to the Included Controls list.
- **Remove controls.** In the Included Controls list, tap the – in a red circle to move a control to the More Controls list.
- **Change the order of the included controls.** Long-press a control's three-line handle, and then drag the control up or down the list.

Summoning Siri

When you first set up the iPhone, you have the option of turning on Siri, during which you can train Siri to your voice by repeating "Hey Siri" and a few other phrases. If you did set up Siri, you're good to go. If you didn't, tap Settings ➪ Siri & Search and set the Press Side Button for Siri switch or the Press Home for Siri switch (depending on your iPhone model) on (green). You can turn off Siri by setting the same switch off (white).

You may also want to choose how Siri responds to you. Choose Settings ⇨ Siri & Search, and then tap Siri Responses to display the Siri Responses screen. Then configure these three settings:

- **Spoken Responses:** Tap the Prefer Silent Responses button, the Automatic button, or the Prefer Spoken Responses button to control when Siri speaks. The Automatic setting lets Siri decide based on its assessment of the situation.
- **Always Show Siri Caption:** Set this switch on (green) if you want Siri's responses to appear onscreen.
- **Always Show Speech:** Set this switch on (green) if you want a transcription of your words to appear onscreen. Some people find this helpful; others don't.

REMEMBER

Having Siri on is a prerequisite to using the Dictation functionality on your phone, discussed later in this chapter. Siri assumes a vital role in searches.

You can call Siri into action in three ways:

- **Press and hold the side button or the Home button.** When your iPhone plays a tone and displays the pulsating Siri icon (see Figure 4-6), start talking.
- **Say "Hey Siri" or "Siri."** On the Siri & Search screen in Settings, tap Listen For to display the Listen For screen. Then tap the "Siri" or "Hey Siri" button or the "Hey Siri" button to tell iOS what to listen for. To turn off this feature, tap the Off button on the Listen For screen.
- **Press a button on a Bluetooth headset or tap an Apple AirPod in your ear.** When you are using a Bluetooth headset or AirPods, you can launch Siri in this way.

FIGURE 4-6: Siri is eager to respond.

Now that Siri is listening, speak your question or command. You can ask a wide range of questions or issue voice commands. See the next section for examples.

What can you say to Siri?

The beauty of Siri is that you don't have to follow a designated protocol when talking to it. Asking, "Will I need an umbrella in New York tomorrow?" produces the same result as "What is the weather forecast around here?" (if *here* was New York City).

Here are examples of ways you can ask Siri for help:

- **Calendar:** "Set up a meeting for 9 a.m. to discuss funding."
- **Clock:** "Wake me up at 8:30 in the morning."
- **Flashlight:** "Turn on Flashlight."
- **Mail:** "Mail the tenant about the rent check."
- **Maps:** "Find an ATM near here."
- **Messages:** "Send a message to Nancy to reschedule lunch."
- **Music:** "Play Frank Sinatra." "Play Beats 1." "What song is this?"
- **Phone:** "Call my wife on her cellphone." "Hey, Siri, hang up." (The other people in the call will hear this.)
- **Photos:** "Show me the pictures I took at Sydney's birthday party." "Show me all the pictures I took at Thanksgiving."
- **Reminders:** "Remind me to take my medicine at 8 a.m. tomorrow."
- **Ride:** "Order me an Uber."
- **Siri Shortcuts:** "Show me my travel plans."
- **Sharing: "**Send this News story (or podcast, song, web page, and so on) to Mary."
- **Smart Home (timed instructions):** "Turn off the bedroom lights in an hour."
- **Sports:** "Who is pitching for the Yankees tonight?"
- **Stocks:** "What is the Dow at?"
- **Trivia:** "Who won the Academy Award for Best Actor in 2003?"
- **Translate:** "How do you say 'where is the bathroom' in Italian?"
- **X (formerly Twitter):** "Send tweet, Going on vacation, smiley-face" or "What is trending on Twitter?"
- **Web search:** "Who was the 19th president of the United States?"

Siri also assumes a more active role in offering search suggestions, contextual reminders, app suggestions, and suggestions based on your location and interests.

Siri shortcuts

Siri includes the potentially helpful Siri Shortcuts feature, whereby Siri can automatically figure out common tasks that may be useful to you or that leverage your calendar or location. You can summon Shortcuts by voice, but not always. Buy coffee every morning on the way to the office, for example, and Siri may display a shortcut on your Lock screen.

You can also create your own shortcuts for those things you do frequently, such as revealing the number of steps you take in the Health app or checking the status of your stocks.

The Shortcuts app contains widgets for shortcuts you've created and a button to create new ones. The Gallery section includes premade shortcuts and widgets, from a speed dial widget to a laundry timer that reminds you when your load is done. You'll also see recommendations for accessibility shortcuts, sharing shortcuts, shortcuts you can run in other apps, and, yes, Siri shortcuts.

Creating your own shortcut involves adding actions found in the app. Say you want to add a shortcut for a workout. One of the available actions you can choose is Log Workout. You must also allow access to the Health app, set a name for the shortcut (such as Workout), and so on. In this example, you can turn on health data categories in the Health app. As a final step, you can record a word or phrase in the Shortcuts app to instruct Siri to kick the shortcut into gear.

Using dictation

In many instances where you'd use the iPhone touchscreen keyboard, you can now use Siri instead. Instead of typing, tap the microphone icon on the keyboard and speak. Tap Done when you're done. Dictation works as you search the web, take notes, compose messages, and so on. You can even update your Facebook status by voice.

Dictation is pretty smart. You can fluidly move between voice and typing because the keyboard stays on the screen while you're dictating. What's more, Dictation now automatically handles punctuation while you dictate. You no longer have to tell it when to insert commas, periods, and question marks, assuming you enunciate and pronounce things clearly. But if you're used to dictating punctuation, carry right on.

Correcting mistakes

Siri's capabilities are impressive, but it doesn't get everything right. Sensibly, Siri seeks your permission before sending a dictated message. If you need to modify the message, you can say "change it" and dictate a new message.

Making Siri smarter

You can tell Siri in which language you want to converse by going to Settings ⇨ Siri & Search. Siri is available in English (United States, United Kingdom, Canada, New Zealand, India, Ireland, Singapore, South Africa, or Australia), Arabic, Chinese, Danish, Dutch, Finnish, French, German, Hebrew, Italian, Japanese, Korean, Malay, Norwegian, Portuguese, Russian, Spanish, Swedish, Thai, and Turkish.

Make sure Siri knows who you are. On the Siri & Search screen in Settings, tap My Information to display your Contacts list, and then tap the card containing the information you want to use. You may want to maintain a separate contact card containing only information you want Siri to know rather than giving Siri your full contact information.

To set Siri's voice, tap Siri Voice on the Siri & Search screen, and then make your choices on the Siri Voice screen. If you're using English (United States) as the language, choose a voice variety in the Variety box — American, Australian, British, Indian, Irish, or South African — and then choose a voice in the Voice box.

Here's a setting you need to make an executive decision about: the Allow Siri When Locked switch on the Siri & Search screen. By default, this switch is set on (green), enabling you — or anyone else — to command Siri to take actions without unlocking the iPhone. If you decide the security threat outweighs the convenience of not having to unlock your iPhone, set this switch off (white).

If you want Siri to integrate with Uber, Venmo, or other third-party apps, look for the respective apps listed under Siri settings, and then tap each app that you want to enable for search, suggestions, shortcuts, or to allow on the Lock screen. You can even ask Siri to learn from the app in question.

Typing to Siri

Speaking to Siri is handy, but you can also summon Siri's assistance by typing. To use this feature, go to Settings ⇨ Accessibility ⇨ Siri, and then set the Type to Siri switch on (green).

Now invoke Siri by long-pressing the side button or the Home button, type your question or command in the space provided, and then tap Done. Siri responds in writing.

IN THIS CHAPTER

» Choosing Apple ID, iCloud, and Family settings

» Configuring networking settings, notifications, and screen time

» Choosing General settings and other key settings

» Specifying Accessibility settings

» Choosing Privacy & Security settings

Chapter 5
Setting Your iPhone Straight

As you see in Chapter 4, Control Center gives you access to the settings you're most likely to change often. When you need to go beyond Control Center's selection of Settings, fire up the Settings app, which lets you reach the full range of settings.

This chapter shows you how to configure most of the widely useful settings — not all settings, because there are too many to cover in a short book. Where other chapters cover particular settings, this chapter points you to the appropriate sections of those chapters.

Meeting the Settings Screen

Tap the Settings icon on the Home screen to display the Settings app. At first, you'll see a long, scrollable list of settings; Figure 5-1 shows the top part of this list. A > symbol appears to the right of most buttons, indicating that you can tap the button to display another screen containing options.

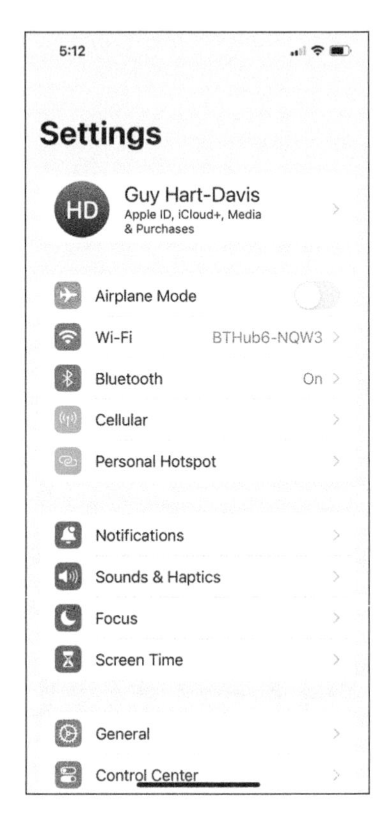

FIGURE 5-1:
The Settings screen is your gateway to the world of iPhone settings.

The settings for iOS and iPhone features appear first, arranged into groups with the items that are typically most frequently used near the top of the screen for quicker access. Settings for built-in apps and apps you've installed appear further down the screen.

Apple ID and iCloud Settings

At the top of the Settings screen is a button bearing your name, or perhaps the mutation of your name preferred by the company that supplied the credit card associated with your Apple ID. I call this button the *Apple ID button.* Tap this button to display the Apple ID screen, where you'll find the following controls:

- **Your picture or monogram (such as your initials):** Tap to display the screen for changing your picture or monogram.
- **Personal Information:** Tap to change your name, birthday, or communication preferences (whether Apple sends you announcements, newsletters, and so on).
- **Sign-In & Security:** Tap to edit the email addresses and phone numbers associated with your account, change your password, and enable two-factor authentication (using another trusted device to authenticate your sign-ins). You can also set up account recover, add a legacy contact who can access your data should you die, and enable Apple's automatic-verification feature to bypass CAPTCHAs.

- **Payment & Shipping:** Tap to manage your payment methods and change your shipping address.
- **Subscriptions:** Tap to manage your subscriptions and enable or disable renewal receipts.
- **iCloud:** Tap to view your iCloud storage status or buy more storage; control which apps use iCloud; manage your iPhone backups and family sharing; configure iCloud+ features if you pay for extra iCloud storage; enable Advanced Data Protection; and control whether you can access your iCloud data on the web.
- **Media & Purchases:** Tap this button, tap View Account, and then sign in to display the Account Settings screen, which includes access to your subscriptions and purchase history.
- **Find My:** Tap to make sure Find My iPhone is enabled and to specify whether your iPhone shares your location.
- **Family Sharing:** Tap to set up Family Sharing or see its status. Family Sharing enables you to share your subscriptions and your purchases of music, movies, and other media with up to five other people you designate as family members. You can invite people aged 13 and over to join the family via Messages, Mail, AirDrop, or other means; for people aged 12 and younger, you can create a child account and introduce the child to it. Once your family members have accepted your invitation, you can use the Family Checklist tool on the Family screen to take actions such as adding your family as emergency contacts, sharing your location, adding recovery contacts in case you lock yourself out of your account, and adding a legacy contact in case you shuffle off this mortal coil.

Also on this screen is a list of devices associated with your Apple ID. Tap a device to see its details; if necessary, tap Remove from Account to remove the device.

At the very bottom of the Apple ID screen is the Sign Out button, which you can tap to sign out of your account. Signing out involves removing all your photos and other iCloud items from the iPhone, so you don't normally want to do this.

Airplane Mode, Wi-Fi, Bluetooth, Cellular, and Personal Hotspot

Below the Apple ID button is a box containing settings for airplane mode, Wi-Fi, Bluetooth, and cellular. Once you've set up the Personal Hotspot feature, Personal Hotspot appears here too.

Airplane mode

Set the Airplane Mode switch on (green) to enable airplane mode. At first, airplane mode turns off all iPhone radios except Bluetooth. You can then turn Bluetooth off manually or turn Wi-Fi on manually; the cellular radio always stays off. When you make these changes, iOS stores them, so the next time you enable airplane mode, you get your customized version.

TIP

When airplane mode is on, the airplane mode icon appears on the status bar, in the upper-right corner on Face ID iPhone models and the upper-left corner on Touch ID models. You can turn airplane mode on or off in Control Center as well as in the Settings app.

Wi-Fi

Wi-Fi is typically the fastest connection for your iPhone, so you'll want to stay connected. Choose Settings ➪ Wi-Fi to display the Wi-Fi screen, and then make sure the Wi-Fi switch at the top is set on (green). You'll then see the Wi-Fi networks that are available (see Figure 5-2).

FIGURE 5-2: Use the Wi-Fi screen in Settings to connect to Wi-Fi networks.

If your iPhone is connected to a Wi-Fi network, that network's name appears below the Wi-Fi switch, with a blue check mark to the left of its name.

The My Networks box shows other available Wi-Fi networks to which your iPhone has previously connected and for which it knows the password. Your iPhone connects automatically to these networks without consulting you, unless you've set the Auto-Join switch off (white) for a particular network. To connect to a different network, tap it.

TIP

To stop your iPhone using one of your My Networks, tap the *i*-in-a-circle next to the network, and then tap Forget This Network.

The Other Networks box shows available networks to which your iPhone has not previously connected (or which you've told it to forget). To connect, tap a network, and then jump through any security hoops that appear, such as typing the password.

The little arcs indicate the signal strength — the more black arcs, the stronger the signal. The lock icons indicate that the network uses security, such as a password you must enter.

At the bottom of the Wi-Fi screen lurk these two settings:

- **Ask to Join Networks:** Tap this button, and then tap Off, Notify, or Ask. Tapping Off means you'll need to join networks in the Other Networks list manually. Tapping Notify makes iOS tell you other networks are available. Tapping Ask makes iOS prompt you to join other networks that are available.
- **Auto-Join Hotspot:** Tap this button, and then tap Never, Ask to Join, or Automatic to control how your iPhone treats available Wi-Fi hotspots.

To reduce the likelihood that your iPhone is being tracked, Apple may serve up a privacy warning if it detects a potential problem with your Wi-Fi network, such as if you're not using a private Wi-Fi address. You can set the Private Wi-Fi Address switch on (green) to use a private address, at least for the networks that support this privacy feature.

Sometimes you may need to connect to a network that doesn't broadcast its name. Such networks are called *hidden* networks or *closed* networks. To connect, tap the Other button in the Other Networks box, and then type the network's name. On the Security screen, set the Private Wi-Fi Address switch on (green) and tap the network's security type — try WPA2/WPA3 first if you don't know which. Type the password and then tap Join.

For safety, avoid connecting to unknown Wi-Fi networks, even if they use security.

Bluetooth

To connect your iPhone wirelessly to nearby electronics, you can use the Bluetooth communications protocol. Bluetooth is great for connecting to headphones, speakers, keyboards, and car systems, among other devices.

To use Bluetooth, you first *pair* the devices, introducing them to each other and establishing a lasting relationship. You can then connect the devices when you want to use them together, and then disconnect and reconnect them as needed. If you want to end the relationship, you *unpair* the devices.

Pair a device with your iPhone

To pair a device, put it into pairing mode by following its instructions. For example, on a speaker, you might hold down its power button until lights start flashing in a particular pattern; on a keyboard, you might press Fn and a specific key.

On your iPhone, choose Settings ⇨ Bluetooth and make sure the Bluetooth switch is set on (green). In the Other Devices list, tap the device, and then follow any prompt to confirm the pairing. Once paired, the device appears in the My Devices box (see Figure 5-3), connected so you can start using it.

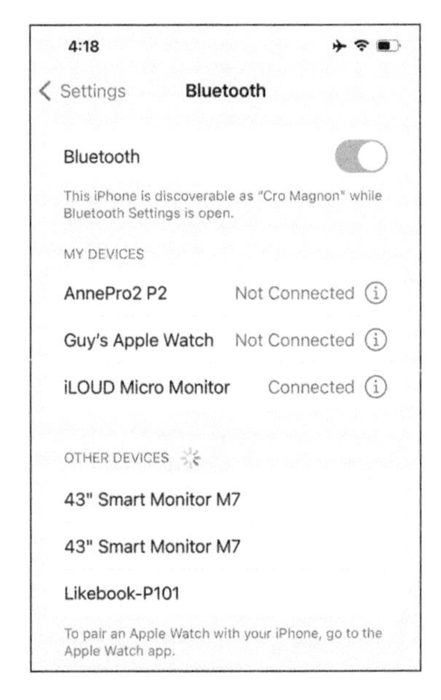

FIGURE 5-3:
The My Devices list on the Bluetooth screen shows paired devices.

WARNING

You can use Bluetooth to share data with certain apps, even when you're not using those apps. To see which apps are using Bluetooth, tap Settings ⇨ Privacy & Security ⇨ Bluetooth. If you want to stop an app from using Bluetooth, set its switch here off (white).

REMEMBER

If you have an Apple Watch, you'll need to open the Apple Watch app on your phone to pair it and handle specific watch-related settings. Apple has also made pairing simpler on some of its other products, including AirPods and Beats headphones.

Disconnect and reconnect a device

To disconnect a connected device, tap the *i*-in-a-circle to the right of the device's name in the My Devices list in the Bluetooth pane, and then tap Disconnect.

To reconnect a paired device, make sure it is powered on, and then tap it in the My Devices list.

Unpair a device

To unpair a device, tap the *i*-in-a-circle to the right of the device's name in the My Devices list in the Bluetooth pane, and then tap Forget This Device. In the confirmation dialog, tap Forget Device.

Cellular

The Cellular category enables you to control your iPhone's cellular connection. The settings vary somewhat depending on your iPhone's model and its carrier. These are the key settings:

- **Cellular Data:** Set this switch on (green) to allow the iPhone to transfer data across the cellular connection.
- **Cellular Data Options:** Tap this button to display the Cellular Data screen. Here, set the Data Roaming switch off (white) to avoid extra charges from connecting to another carrier's network when outside your regular carrier's area. Tap Voice & Data to specify the connection technology to use, such as 5G Auto, 5G On, or 4G. Tap Data mode and select the data mode, such as Allow More Data on 5G, Standard, or Low Data Mode. Set the Limit IP Address Tracking switch on (green) to hide your iPhone's IP address from known trackers when you're using Mail and Safari.
- **Personal Hotspot:** If your carrier plan includes sharing your iPhone's internet connection, tap this button and choose settings on the Personal Hotspot screen. Set the Allow Others to Join switch on (green) to enable others to join the hotspot, and then type in the Wi-Fi password field the password they must enter to join. If you use Family Sharing, tap Family Sharing and specify which family members can join automatically and which need your approval. Set the Maximize Compatibility switch on (green) if PCs running Windows or Linux or devices running Android will need to connect.
- **Carrier Services:** Tap this button to display information about your carrier, such as phone numbers for contacting them.
- **Wi-Fi Calling:** Tap this button to enable or disable phone calls via Wi-Fi connections if your carrier supports them. Wi-Fi calling can help you avoid running over your cellular call allowance and racking up extra fees.
- **Calls on Other Devices:** Tap this button to specify which of your non-iPhone Apple devices, such as iPads and Macs, can make phone calls via your iPhone. Apple calls this feature Continuity. Each device must be signed into the same iCloud account.
- **Cellular Data box:** The Current Period readout shows the amount of data you've used in your current billing period. The Current Period Roaming readout shows roaming usage. The apps list shows which apps are allowed to

use cellular data and how much they've used. If the list is headed Apps by Usage, you can tap Sort by Name to switch to Apps by Name order, and tap Sort by Usage to switch back. Typically, each app's switch is set initially to the on position; set the switch off (white) for each app that you don't want using cellular data.

» **Wi-Fi Assist:** Set this switch on (green) if you want your iPhone to automatically use cellular data when its Wi-Fi connection is flaky. This feature can chew through your data allowance, so set the switch off (white) unless your allowance is generous or unlimited.

» **Call Time:** The readouts let you know how many hours and minutes you've spent on the phone for the current period and for all the time you've used this iPhone.

Notifications, Sounds & Haptics, Focus, and Screen Time Settings

The next group of settings lets you control notifications; sounds and haptic feedback; focus (do not disturb); and Screen Time, Apple's self-control and parental-controls feature.

Notifications

Your iPhone displays important notifications as they arrive and collects them as well as less exciting notifications in Notification Center, which you can open by swiping down from the top of the screen.

To control which notifications appear where, choose Settings ➪ Notifications and work on the Notifications screen (see Figure 5-4). Here, you can configure the following settings:

» **Display As:** In this box, tap Count, Stack, or List to set the default means for displaying notifications. You can then change them for individual apps, as needed. Count shows only the number of notifications waiting for you; Stack shows the notifications stacked on top of one another; and List shows the notifications spread out so that you can scan them quickly.

» **Scheduled Summary:** To create a scheduled summary of your notifications, tap this button, and then set the Scheduled Summary switch on (green). Follow through the resulting screens to choose which apps to include in the summary and when to display the summary. The default schedule suggests one summary at 8:00 AM and one at 6:00 PM, but you can change the times, delete the second summary, or add further summaries, as needed. The summaries then appear on the Scheduled Summary screen, where you can change them if you want.

- **Show Previews:** To set your default setting for showing previews, tap this button, and then tap Always, When Unlocked, or Never. Be warned that Always may expose sensitive previews to other people who can see your phone when it is locked. You can choose different preview settings for individual apps.
- **Screen Sharing:** To control whether notifications appear while you're sharing your iPhone's screen, tap this button, and then set the Allow Notifications switch on (green) or off (white).
- **Siri Suggestions:** To specify which apps can suggest shortcuts on the Lock screen, tap this button, set the Allow Notifications switch on (green), and then set the switch for each individual app on (green) or off (white).
- **Notification Style:** In this box, tap the app you want to configure, and choose settings on the screen that appears. Set the Allow Notifications switch off (white) if you want to suppress notifications. Otherwise, leave the switch on (green) and configure notification preferences. The most important are choosing between Immediate Delivery and Scheduled Delivery in the Notification Delivery box; selecting the alert style in the Alerts box; selecting either Temporary or Persistent for the Banner Style; and setting the Badges switch to control whether the app's icon on the home screen can display a red badge containing the number of notifications.
- **Government Alerts/Emergency Alerts:** At the bottom of the Notifications screen, examine the selection of controls for receiving emergency alerts, and then set each switch on (green) or off (white), as needed. The box name and the alerts it contains vary by country and region.

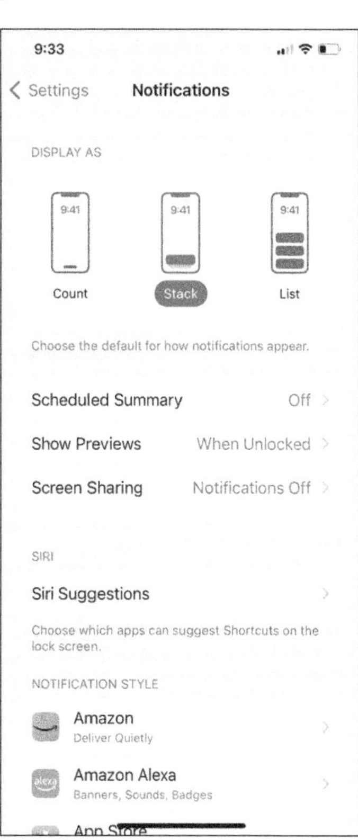

FIGURE 5-4:
Notify the iPhone of your notification intentions.

Sounds & Haptics

Haptics is technology that stimulates the senses of touch and motion, such as when your iPhone vibrates to give you feedback. To configure how your iPhone plays sounds and gives you haptic feedback, choose Settings ➪ Sounds & Haptics, and then work on the Sounds & Haptics screen. The following list explains the main settings:

- **Headphone Safety:** Tap this button to display the Headphone Safety screen. Here, you can set the Reduce Loud Sounds switch on (green) and then drag the slider to a suitable level to protect your hearing.
- **Personalize Spatial Audio:** Your iPhone can photograph your ears and use the information it gleans to create a personalized spatial audio profile for you for supported AirPods or Beats headphones. Tap this button, tap Personalized Spatial Audio, and follow the prompts to take aural portraits.
- **Ringtone and Alert Volume:** Drag the slider to set the volume. Set the Change with Buttons switch on (green) to enable changing the volume by pressing the volume up and volume down buttons.
- **Sounds and Haptic Patterns:** In this box, tap each button — Ringtone, Text Tone, and so on — in turn, and then specify the tone and vibration to use.
- **Keyboard Feedback:** Tap this button, and then set the Sound switch and the Haptic switch on (green) or off (white), as needed. Setting Sound on plays typewriter noises.
- **Lock Sound:** Set this switch on (green) to play a sound when your iPhone locks.
- **Play Haptics in Ring Mode:** Set this switch on (green) to play haptics while your phone is ringing aloud.
- **Play Haptics in Silent Mode:** Set this switch on (green) to play haptics while your phone is in silent mode.
- **System Haptics:** Set this switch on (green) to play haptics when your iPhone needs your attention.

Focus

Your iPhone's Focus feature can help you maintain concentration for most any activity. To configure Focus, tap Settings ➪ Focus, and then work on the Focus screen (see Figure 5-5). Your iPhone includes several built-in focuses, such as Do Not Disturb, Driving, and Work. You can create other focuses, as needed.

To configure a focus that appears in the box at the top of the screen, tap it, and then work on the configuration screen, such as the Do Not Disturb screen. You can:

- » Choose which people and which apps can interrupt the focus with notifications.
- » Choose whether silenced notifications appear on the Lock screen (tap Options).
- » Add a custom Lock screen or Home screen to the focus to minimize distractions.
- » Set one or more schedules for the focus.
- » Apply a focus filter, such as limiting the Mail app to displaying only specific inboxes or restricting Safari to a particular profile or tab group.

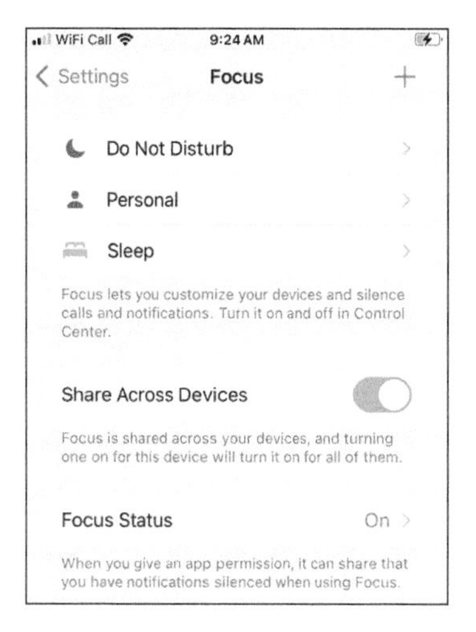

FIGURE 5-5: Set up a focus for each activity that requires your undivided attention.

To configure a focus that doesn't appear in the box, tap + (add) on the Focus screen and follow the prompts.

To turn on a focus, open Control Center, tap the Focus text on the Focus button (don't tap the icon), and then tap the focus in the list. To set the duration for any focus but Driving, tap the ellipsis (. . .) button, and then tap a button, such as For 1 Hour, Until This Evening, or Until I Leave This Location. The focus's icon appears in the status bar. To turn off the focus, open Control Center and tap the Focus icon.

Your iPhone automatically activates the driving focus when it detects you in apparently vehicular motion.

You can grant permission to certain apps to share your focus status. For example, you might want your Messages contacts to know that you're focusing rather than ghosting them.

Screen Time

Screen Time combines parental controls (you know, for controlling your kids — or your parents) with the capability to police your own behavior. If you enable all its features, Screen Time enables you to see everything from which apps you use most, how many times you pick up your iPhone in a day, and how many notifications you receive.

To get started, choose Settings ⇨ Screen Time. On the Screen Time screen (shown on the left in Figure 5-6), work with the following controls to specify the restrictions you need:

- **App & Website Activity:** To set up the App & Website Activity feature, tap this button, and then tap Turn On App & Website Activity. The Daily Average histogram appears at the top of the Screen Time screen, showing your iPhone usage. Tap See All App & Website Activity to display a screen that shows histograms for Screen Time, Pickups, and Notifications; you can then tap the Week tab or the Day tab at the top of the screen to change the time period. When you want to turn off App & Website Activity, go to the bottom of the Screen Time screen and tap the Turn Off App & Website Activity button.
- **Downtime:** To enforce a downtime schedule of time away from the screen, tap Downtime on the Screen Time screen. You can then either tap Turn On Downtime Until Tomorrow or set the Scheduled switch on (green) to specify a schedule. The schedule can use either the same hours for each day or custom hours for different days.
- **App Limits:** To apply time limits for different app categories or individual apps, tap App Limits. On the App Limits screen, tap Add Limit, and then select the app category, such as the Social Category or the Games category, or select one or more individual apps within a category. Tap Next, set the time limit, and then tap Add.
- **Always Allowed:** To specify contacts and apps that Screen Time always allows, tap Always Allowed. On the Always Allowed screen, tap Contacts, and then specify the allowed contacts. Then customize the Allowed Apps list by tapping the remove icon (red circle containing –) to remove an existing item or tapping the add icon (green circle containing +) to add an item from the Choose Apps list. The Phone app is always allowed.
- **Screen Distance:** To have the iPhone display alerts when the user is holding it too close to their face, tap Screen Distance, and then set the Screen Distance switch on (green).
- **Communication Limits:** To restrict the contacts with whom the user may communicate, tap Communication Limits. On the Communication Limits screen, tap During Screen Time and tap Contacts Only, Contacts & Groups with At Least One Contact, or Everyone; then tap During Downtime and tap Specific Contacts (whom you then specify) or Everyone, as needed.
- **Communication Safety:** To help avoid children sharing nude photos and videos, tap Communication Safety, and then set the Check for Sensitive Photos switch on (green).

» **Content & Privacy Restrictions:** To limit the actions the user can take, tap Content & Privacy Restrictions. On the Content & Privacy Restrictions screen (shown on the right in Figure 5-6), set the Content & Privacy Restrictions switch on (green), and then use the controls to specify which features to allow and which to block. For example, to restrict movies, tap Content Restrictions, tap Movies, and then tap the maximum rating you'll allow, such as PG-13.

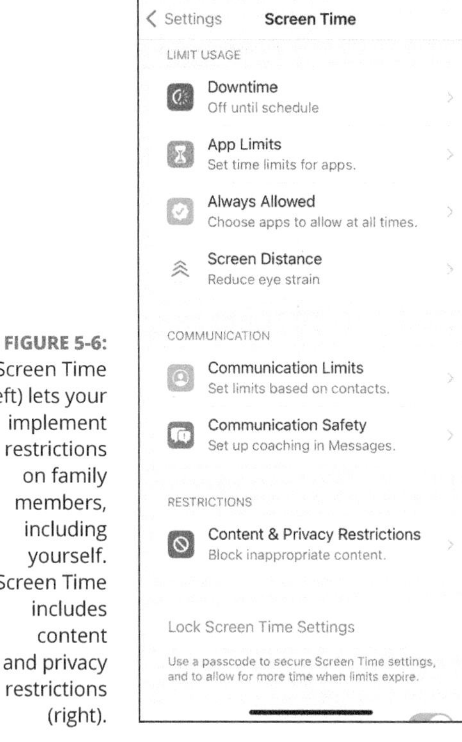

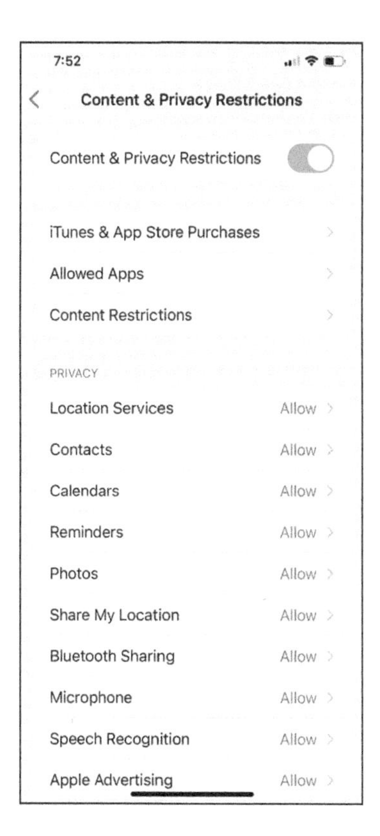

FIGURE 5-6: Screen Time (left) lets your implement restrictions on family members, including yourself. Screen Time includes content and privacy restrictions (right).

» **Lock Screen Time Settings:** Tap this button to lock Screen Time settings with a passcode.

» **Share Across Devices:** Set this switch on (green) to share this user's Screen Time settings across all the devices linked to their Apple ID.

» **Set Up Screen Time for Family:** Tap this button to set up Screen Time for members of your Family group.

General Settings

The General screen (see Figure 5-7) gives you access to a range of settings screens that run the gamut from information about your iPhone to commands for resetting it or shutting it down.

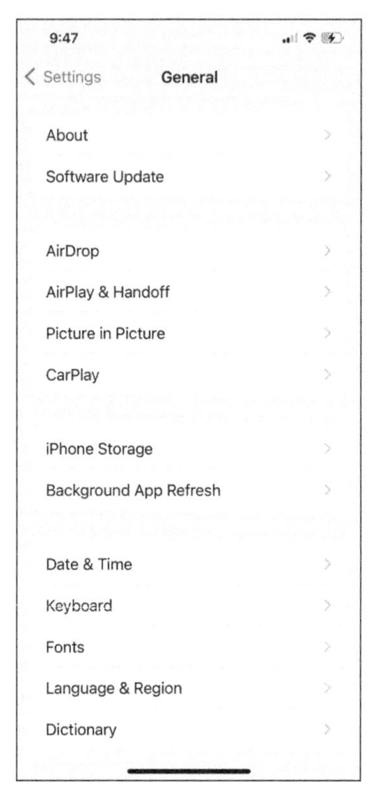

FIGURE 5-7:
The General screen in Settings.

About

The About screen shows a slew of information about your iPhone. Tap the Name button to change the iPhone's display name. Look at the iOS Version readout, Model Name readout, Model Number readout, and Serial Number readout to get essential details for troubleshooting and warranty queries. Other key pieces of information include the Wi-Fi address and Bluetooth address readouts.

Software Update

Tap Software Update to see and install any available update. To configure updating, tap Automatic Updates. For the best performance and highest security, go to the Automatically Install box and set the iOS Updates switch and Security Responses & System Files switch on (green). Then go to the Automatically Download box and set the iOS Updates switch on (green).

AirDrop

Tap AirDrop to set your iPhone's status for AirDrop file transfer: Receiving Off, Contacts Only, or Everyone for 10 Minutes. You can change AirDrop status more quickly in Control Center.

NEW

In the Start Sharing By box, set the Bringing Devices Together switch on (green) if you want to trigger the NameDrop sharing feature by bringing your iPhone close to another iPhone or an iPad.

AirPlay and Handoff

Apple's Handoff feature enables you to pass tasks between devices signed in to your iCloud account. For example, you might start an email on your Mac, resume it on your iPhone on the bus, and then complete it on your iPad at the destination. On your iPhone or iPad, you resume the task from the Lock screen or the app switcher. On your Mac, you resume the task from the dock. Set the Handoff switch on (green) to use Handoff.

To control whether your iPhone automatically connects to AirPlay speakers, tap Automatically AirPlay, and then tap Never, Ask, or Automatic.

Set the Transfer to HomePod switch on (green) if you have an Apple HomePod speaker and want your iPhone to start playing via the HomePod when you bring the iPhone close to the top of the HomePod.

Set the Continuity Camera switch on (green) to use your iPhone's cameras as your Mac's webcam.

Picture in Picture

To have videos and FaceTime calls switch to picture-in-picture (PIP), showing a miniature picture but still playing, when you display the Home screen or switch to other apps, tap the Picture-in-Picture button, and then set the Start PIP Automatically switch on (green).

CarPlay

Tap the CarPlay button to display its screen, where you can set up CarPlay to work with your CarPlay-compliant vehicle. Settings include choosing whether to allow CarPlay to work when the phone is locked and customizing the CarPlay apps.

Home Button

On the iPhone SE, tap the Home button to display the Home Button Haptic screen, which enables you to choose among three haptic feedback types when you press the Home button. Tap the 1 circle, the 2 circle, or the 3 circle, and then press the Home button to see how that haptic type feels. When you find your favorite, tap Done.

iPhone Storage

Tap the iPhone Storage button to display the iPhone Storage screen (see Figure 5-8), which enables you to review and manage your iPhone's storage. The histogram at the top shows how much space is free and what types of files are occupying the rest.

The Recommendations box contains suggestions for freeing up space, such as by offloading unused apps and deleting large attachments. The large unnamed box below the Recommendations box contains apps, initially sorted by size; to change the sort, tap the current sort method, and then tap Size, Name, or Last Used Date. Tap an app to display its information screen, from which you can offload the app (deleting the app but keeping its data) or delete both the app and its data. Look at the App Size readout and the Documents & Data readout to inform your decision.

At the bottom of the screen, the last box displays an iOS readout and a System Data readout showing how much space the operating system and its system data are taking up.

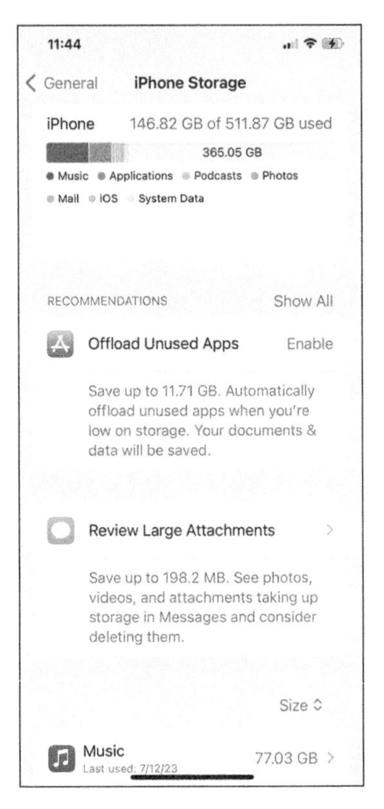

FIGURE 5-8: Review and manage your iPhone's storage on the iPhone Storage screen.

Background App Refresh

Tap the Background App Refresh button, and then choose settings to control whether apps other than the foreground app can refresh data via Wi-Fi or cellular connections. Tap Background App Refresh, and then tap Off, Wi-Fi, or Wi-Fi & Cellular Data, as needed. Normally, you'd choose Wi-Fi unless you have a generous or unlimited cellular plan. Then set each app's switch on (green) or off (white) to specify which can refresh in the background.

Date & Time

Tap Date & Time to set the date and time manually if necessary, to enable or disable 24-hour time, and to set the time zone manually.

Keyboard

Tap Keyboard to display the Keyboards screen, where you can

- Tap Keyboards to manage software keyboards.
- Tap Hardware Keyboard, when one is connected, to choose settings for it.
- Tap Text Replacement to manage text replacements (such as AutoCorrect).
- Tap One-Handed Keyboard to enable the left-handed or right-handed one-hand keyboard.
- Choose text-input settings, such as Auto-Capitalization, Auto-Correction, and Check Spelling.
- Enable or disable Dictation and Auto-Punctuation.

Fonts

Tap Fonts to manage fonts you've installed from the App Store.

Language & Region

Tap Language & Region to set your language, such as English, and region, such as United States. You can choose the calendar type, temperature system, the first day of the week, and more. Set the Live Text switch on (green) if you want to capture text using your iPhone's camera.

Dictionary

Tap Dictionary to choose which dictionaries are available on your iPhone. A huge list of dictionaries is available, most of which download when you tap them.

VPN & Device Management

A *virtual private network*, or *VPN*, uses encryption to create a secure "tunnel" across the insecure internet to a VPN server. You'd typically use a VPN either to connect securely to your company's network from a remote location or to avoid geographical restrictions on content — for example, making your iPhone appear to be in the U.S. when you are traveling.

You can set up a VPN configuration manually (start by tapping VPN & Device Management), but these days it's more likely you can use one of the following approaches:

- **VPN profile:** A network administrator provides a profile that you install on your iPhone, setting up the VPN automatically.
- **VPN app:** Many VPN providers supply apps that handle the complexities of the connections for you. You supply your credentials, and the VPN works.

The VPN & Device Management screen also enables you to see any configuration profiles applied to your iPhone for remote management.

Control Center Settings

Tap Control Center Settings to configure Control Center. Set the Access within Apps switch on (green) to enable yourself to open Control Center while an app is active. Set the Show Home Controls switch on (green) if you want Control Center to include controls for Home app accessories. Choose other controls for Control Center by tapping the remove icon (red circle containing –) in the Included Controls List, tapping the add icon (green circle containing +) in the More Controls list, and then dragging the controls in the Included Controls list into your preferred order.

See Chapter 4 for information on using Control Center.

Display & Brightness Settings

Tap Display & Brightness to show its screen, on which you can configure the following settings:

- **Appearance:** In this box, tap Light or Dark to set the overall appearance. To have the iPhone switch between light and dark, set the Automatic switch on (green), tap Options, and then either tap Sunset to Sunrise or tap Custom Schedule and specify the schedule.
- **Text Size:** Tap to set the text size for the many apps that support dynamic type.
- **Bold Text:** Set this switch on (green) to use a bold typeface.

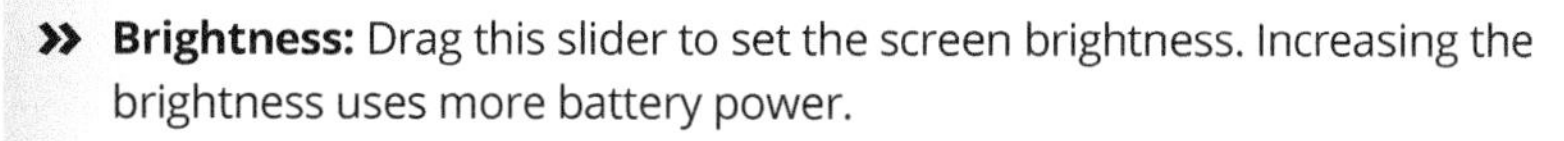

- **Brightness:** Drag this slider to set the screen brightness. Increasing the brightness uses more battery power.

TIP

To have the iPhone set screen brightness automatically, choose Settings ⇨ Accessibility ⇨ Display & Text Size, scroll to the bottom, and then set the Auto-Brightness switch on (green).

- **True Tone:** Set this switch on (green) to have the iPhone make the colors appear consistent in different lighting conditions.
- **Night Shift:** Tap to configure the night shift feature, which automatically alters the colors on your iPhone screen after dark so that they're at the warmer end of the color spectrum to reduce disruptions to your sleep. To use night shift on a schedule, set the Scheduled switch on (green), tap the From/To button, and specify the schedule. Otherwise, turn on night shift as needed by setting the Manually Enable Until Tomorrow switch on (green). Either way, drag the Color Temperature slider along the Less Warm–More Warm Axis to set the color temperature for night shift to use.
- **Auto-Lock:** Tap to set the time that elapses before the iPhone automatically locks and turns off the display. Your choices are 30 Seconds, 1 Minute, 2 Minutes, 3 Minutes, 4 Minutes, 5 Minutes, or Never.

TIP

If your iPhone is managed via corporate policy, the Never item doesn't appear on the Auto-Lock screen.

- **Raise to Wake:** Set this switch on (green) to have your iPhone wake up when you raise it to your face.
- **Display Zoom:** To switch between the default text size and the larger size, tap this button, tap Larger Text or Default Text, and then tap Done.

Home Screen & App Library Settings

Tap the Home Screen & App Library button to choose the following settings:

- **Newly Downloaded Apps:** in this box, tap either Add to Home Screen or App Library Only, as needed.
- **Show in App Library:** Set this switch on (green) to display badges in App Library. Badges are the red rounded rectangles containing the number of updates.
- **Show on Home Screen:** Set this switch on (green) if you want to display the Search box on the Home screen in place of the row of dots indicating which Home screen page is active.

Accessibility Settings

iOS includes a huge number of accessibility features aimed at helping people with various disabilities use the iPhone effectively. The Accessibility screen divides the features into five boxes: Vision, Physical and Motor, Hearing, Speech, and General. The following subsections explain the features you'll find in each box.

Accessibility features in the Vision Box

The Vision box contains the following features:

- **VoiceOver:** Tap to configure the VoiceOver screen reader, which announces what is displayed on screen. Be warned that touchscreen gestures change when you set the VoiceOver switch on (green). For example, you tap to select an item, and then double-tap to activate it.
- **Zoom:** Tap to configure and enable this screen magnifier. With zoom enabled, you double-tap the screen with three fingers to zoom in, drag with three fingers to move around the screen, and double-tap with three fingers to zoom out again.
- **Display & Text Size:** Tap to enable features for making the display and text easier to see. These include Bold Text, Larger Text, Button Shapes (underlines appear on words that are buttons), and On/Off labels on switches. Other settings let you reduce transparency in the interface, increase contrast, differentiate UI controls without color, invert either all colors (Classic Invert) or all colors except those in images and media (Smart Invert), apply color filters for color blindness, and reduce the white point.

TIP

At the bottom of the Display & Text Size screen lurks the Auto-Brightness switch. When this switch is on (green), as it is by default when you first set up your iPhone, iOS adjusts the screen brightness to match the ambient lighting conditions. Many people find Auto-Brightness makes the screen too dim. Setting the Auto-Brightness switch off (white) makes the screen look better at the expense of battery life.

- **Motion:** Tap to display the Motion screen, where you can set five switches on (green) or off (white) to decrease motion effects: Reduce Motion (to reduce the parallax effect that makes icons float above the Home screen), Auto-Play Message Effects, Animated Images, Dim Flashing Lights, and Auto-Play Video Previews.
- **Spoken Content:** Set the Speak Selection switch on (green) to make iOS display a Speak button when you select text; tap Speak to have the text spoken. Set the Speak Screen switch on (green) to have iOS read the screen's

content when you swipe down with three fingers from the top. You can also configure typing feedback, choose the speaking voice, set the speaking rate, and set up custom pronunciations.

» **Audio Descriptions:** On the Audio Descriptions screen, set the Audio Descriptions switch on (green) to have iOS play audio descriptions automatically.

Accessibility features in the Physical and Motor box

The Physical and Motor box contains the following features:

» **Touch:** On the Touch screen, configure AssistiveTouch if you have difficulty tapping the screen accurately. Set the Reachability switch on (green) to use Reachability (see Chapter 2). Tap Haptic Touch to change the touch sensitivity. Tap Touch Accommodations to change how the screen responds to touches. Set the Tap to Wake switch on (green) to wake the screen by tapping. Set the Shake to Undo switch off (white) if you find yourself undoing actions unintentionally by shaking your iPhone. Set the Vibration switch off (white) if you need to disable all vibrations (including emergency alerts). Set the Prevent Lock to End Call switch on (green) if you find yourself ending calls by accident by pressing the side button. Tap Call Audio Routing and choose how to route calls. Tap the Back Tap button and then choose actions to take when you double-tap or triple-tap the back of the iPhone.

» **Face ID & Attention:** On this screen, set the Require Attention for Face ID switch on (green) to prevent your family from unlocking your iPhone by holding it in front of your sleeping face. Set the Attention Aware Features switch on (green) to have your iPhone check for your attention before taking actions such as dimming the screen. Set the Haptic on Successful Authentication switch on (green) to receive haptic feedback when Face ID unlocks your iPhone or approves a purchase.

» **Switch Control:** On this screen, you can set up your iPhone to take actions when you use switches. These can be either external physical switches, such as buttons or joysticks, or virtual switches you set up using the iPhone's hardware controls and accelerometers.

» **Voice Control:** On this screen, you can set up your iPhone so you can control it using your voice. iOS includes many built-in commands, but you can create custom commands, as needed.

» **Side Button:** On this screen, set the side button's click speed (Default, Slow, or Slowest) and the action to take when you press and hold the button. You can also set confirmation for use of the side button with Switch Control and AssistiveTouch.

- **Control Nearby Devices:** From this screen, you can start setting up control of nearby devices signed into the same iCloud account.
- **Apple TV Remote:** If you're using your iPhone as a remote control for Apple TV, set the Directional Buttons switch on (green) to use buttons on the iPhone rather than swipe gestures. Set the Live TV Buttons switch on (green) to always display guide and channel buttons.
- **Keyboards:** On this screen, you can set up Full Keyboard Access, enabling you to control your iPhone using a hardware keyboard.

Accessibility features in the Hearing box

The Hearing box contains the following features:

- **Hearing Devices:** On this screen, you can manage hearing devices that meet the Made for iPhone (MFI) standard. Set the Hearing Aid Compatibility switch on (green) to improve audio quality with some hearing aids.
- **Sound Recognition:** You can have your iPhone listen for particular noises, such as fire alarms, dogs barking, or the door bell ringing. On the Sound Recognition screen, set the Sound Recognition switch on (green), then tap Sounds. On the Sounds screen, tap the sound you want to hear, set its switch on (green), and then tap Alert Tones and select the tone to play.
- **RTT/TTY:** (Some carriers only.) Real-Time Text (RTT) lets your iPhone send text as you create it, without you having to tap a Send button. This feature can be good for conversational flow, but it's not great if you prefer to edit your texts before sending them. Teletypewriter (TTY) enables your iPhone to send text messages over a phone connection.
- **Audio/Visual:** On this screen, tap Headphone Accommodations to set up customized audio for Apple and Beats headphones that support it. Tap Background Sounds to set up a background sound, such as Rain or Dark Noise, to block out unwanted environmental noise. Set the Mono Audio switch on (green) if you want mono audio output instead of stereo. Set the Phone Noise Cancellation switch on (green) to reduce ambient noise on phone calls when you're holding the iPhone to your ear. Set the Headphone Notifications switch on (green) to have your iPhone notify you when you've been listening to loud audio long enough to affect your hearing. Drag the Balance slider to set the left-right balance on audio output. Tap LED Flash for Alerts to configure flashing to notify you when alerts arrive.

- **Subtitles & Captioning:** On this screen, set the Closed Captions + SDH switch on (green) to use closed captions and subtitles for the deaf and hard of hearing (SDH). Tap Style and choose the style for closed captions and subtitles. Set the Show Audio Transcriptions switch on (green) to show audio transcriptions for announcements from your HomePod.
- **Live Captions:** On this screen, set the Live Captions switch on (green) to have iOS listen to media content you're playing and attempt to display live captions of the words. Tap Appearance and configure the appearance for the captions. Set the Live Captions in FaceTime switch on (green) if you want to use live captions in FaceTime.
- **Hearing Control Center:** On this screen, you can manage your hearing controls and drag them into your preferred order.

Accessibility features in the Speech box

The Speech box contains either one or two features, depending on your iPhone model:

- **Live Speech:** On this screen, set the Live Speech switch on (green) to enable Live Speech, which speaks aloud text you type or phrases you select from a Favorite Phrases list. Tap Favorite Phrases and set up your list of phrases you want to be able to use quickly. Tap Voices and select the voice that will speak. You can then launch Live Speech by using the Accessibility Shortcut (discussed in just a moment).

NEW

- **Personal Voice:** (More powerful iPhone models only.) On this screen, tap the Create a Personal Voice button and follow the prompts to record 150 voice snippets from which iOS can create a digital facsimile of your voice. You can then use this personal voice with the Live Speech feature to read screen content aloud, either to yourself or when communicating with others. Set the Share across Devices switch on (green) if you want to use Personal Voice on your other iCloud devices. Set the Allow Apps to Request to Use switch on (green) to allow apps to request to use Personal Voice.

Accessibility features in the General box

The General box contains the following features:

- **Guided Access:** The Guided Access feature enables you to lock the iPhone into a single app, protected by a passcode, and set time limits for usage. You might want to use Guided Access to let someone you don't trust use your iPhone to accomplish a particular task, without risking them accessing your other apps or data. Guided Access is also useful in settings such as education and training, but typically on iPads rather than iPhones.

- **Assistive Access:** The Assistive Access feature lets you switch the iPhone to use a simplified user interface with larger controls and fewer apps. You would typically set up Assistive Access for someone else rather than use it yourself.
- **Siri:** On the Siri screen, set the Type to Siri switch on (green) if you want to be able to communicate with Siri by typing instead of by speaking. In the Siri Pause Time box, set the time Siri waits for you to finish speaking by tapping Default, Longer, or Longest. Drag the Speaking Rate slider along the tortoise–hare axis to change Siri's speaking rate. In the Spoken Responses box, control how much Siri speaks by tapping Prefer Silent Responses, Automatic, or Prefer Spoken Responses. Set the Always Listen for "Hey Siri" switch on (green) if you want Siri to listen even when your iPhone is facedown or covered. Set the Show Apps behind Siri switch on (green) to keep the current app visible while Siri is active. Set the Announce Notifications on Speaker switch on (green) to have Siri announce your notifications aloud through your iPhone's speaker.
- **Accessibility Shortcut:** When you triple-click the side button on a Face ID iPhone or triple-click the Home button on a Touch ID iPhone, iOS runs the Accessibility Shortcut, which gives you quick access to one or more Accessibility features. If you assign a single feature to the Accessibility Shortcut, triple-clicking triggers that feature. If you assign multiple features to the Accessibility Shortcut, iOS displays the Accessibility Shortcuts dialog (see Figure 5-9), in which you tap the feature you want this time. To specify your Accessibility Shortcuts, select one or more items on the Accessibility Shortcut screen; if you select multiple items, drag them into the order in which you want them to appear in the dialog.

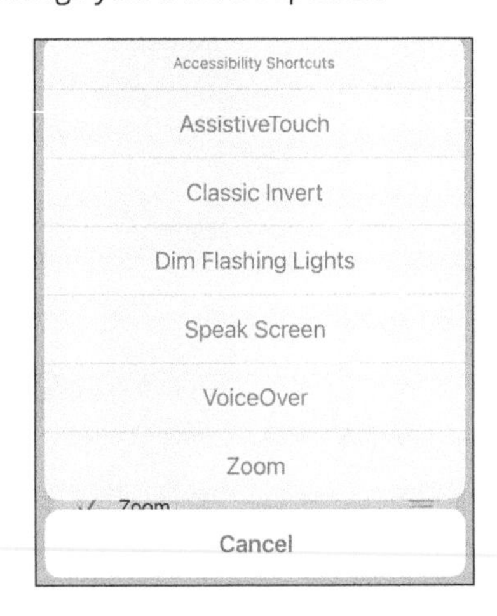

FIGURE 5-9: The Accessibility Shortcuts dialog gives you quick access to your chosen Accessibility features.

- **Per-App Settings:** iOS enables you to configure various vision features and motion features for individual apps. For example, you could make the App Store app appear with bold text, smartly inverted colors, and reduced motion without applying these settings to all apps. On the Per-App Settings screen, tap Add App, and select the app. Then tap the app's name to display the customization settings.

Wallpaper

Tap the Wallpaper button to select wallpaper for your iPhone's Lock screen and Home screen. Tap Add New Wallpaper to display the Add New Wallpaper screen (shown on the left in Figure 5-10), and then make a selection from the buttons across the top (Photos, Photo Shuffle, Emoji, Weather, Astronomy, or Color) or from the categories down the screen (Featured, Photo Shuffle, Weather & Astronomy, Kaleidoscope, Emoji, Unity, Pride, Collections, and Color). On the preview screen that appears (shown on the right in Figure 5-10), tap Add Widgets to add any widgets you want the wallpaper to display.

When you finish, tap Add in the upper-right corner. In the dialog that appears, tap Set as Wallpaper Pair to use the wallpaper for both the Home screen and the Lock screen or tap Customize Home Screen to use it only for the Home screen.

TIP

You can also change the wallpaper directly from the Lock screen. See Chapter 2 for details.

FIGURE 5-10: On the Add New Wallpaper screen (left), tap the wallpaper you want. On the preview screen (right), add any widgets you'd like.

Standby

StandBy is a new feature in iOS 17 that displays widgets, photos, or clocks on the screen when the iPhone is charging and turned on its side. Set the StandBy switch on (green) to enable StandBy. Set the Night Mode switch on (green) if you want StandBy to appear with a red tint. Set the Show Notifications switch on (green) if you want to see all notifications; critical notifications will appear even if you set this switch off (white).

Siri & Search

On the Siri & Search screen, choose when and how to use Siri and which suggestions to allow. See Chapter 4 for coverage of the key settings.

Face ID/Touch ID & Passcode

On the Face ID/Touch ID & Passcode screen, choose when to use Face ID or Touch ID (depending on your iPhone model):

» **Use Face ID/Touch ID For:** In this box, set the iPhone Unlock switch, the iTunes & App Store switch, the Wallet & Apple Pay switch, and the Password AutoFill switch on (green) or off (white), as needed.

» **Set Up an Alternate Appearance:** Tap this button to teach Face ID to recognize you with an alternate appearance, such as when you've applied full-face make-up.

» **Face ID with a Mask:** If this switch appears, set it on (green) to allow yourself to use Face ID while wearing a protective mask (not a Halloween mask). As of this writing, this feature is more well-intentioned than successful.

» **Add Glasses:** If you've set up Face ID with a mask, tap this button to add glasses to the mix. You can't add shades.

» **Reset Face ID:** Tap this button to reset Face ID so that you can set it up again from scratch.

» **Require Attention for Face ID:** Set this switch on (green) to have Face ID make sure you're looking at the iPhone to unlock it.

- **Attention Aware Features:** Set this switch on (green) to have your iPhone check for your attention before taking actions such as dimming the screen or expanding a notification on the Lock screen.
- **Finger 1 (or other name):** Tap the finger's button to display a screen on which you can rename the button more descriptively or delete the fingerprint details.
- **Add a Fingerprint:** For Touch ID, tap this button to add another fingerprint. You can add up to five fingerprints.
- **Turn Passcode Off:** If for some unfathomable reason you need to turn off the passcode, tap this button, and then tap Turn Off in the Turn Off Passcode? dialog that opens. Be warned that when you remove the passcode, iOS strikes back by removing your Apple Pay cards and your car keys from Wallet and prevents your Apple Watch from unlocking automatically.
- **Change Passcode:** Tap this button to start changing your passcode.
- **Expire Previous Passcode Now:** If this button is available, you can tap it to make your previous passcode expire.
- **Voice Dial:** Set this switch on (green) to allow dialing by voice.
- **Allow Access When Locked:** In this box, set the switches on (green) or off (white) to specify which apps and features you will and will not allow while the iPhone is locked.

Allowing access while your iPhone is locked is convenient, but it becomes a security risk if someone else gets their hands on your iPhone.

- **Erase Data:** Set this switch on (green) to make your iPhone erase all your data after ten successive failed attempts to enter the password. This is a strong security measure against brute-force breaking of the password, but bear in mind that it gives a malefactor an easy way to erase your valuable data. To delay such an attack, iOS implements an increasing timeout between each login attempt after the first few.

Emergency SOS

On the Emergency SOS screen, choose settings for making emergency calls:

- **Call with Hold and Release:** Set this switch on (green) to call by holding down the side button and either the volume up button or the volume down button for several seconds. The Power Off screen appears and a countdown begins; when the countdown ends, release the buttons to call emergency services.

- **Call with 5 Button Presses:** Set this switch on (green) to start an emergency call by pressing the side button five times in quick succession. The Emergency SOS screen appears, and a 10-second countdown starts; when the countdown finishes, the iPhone places the emergency call.
- **Call Quietly:** Set this switch on (green) to suppress alarms, flashes, and VoiceOver audio when calling with either Call with Hold and Release or Call with 5 Button Presses.
- **Call after Severe Crash:** Set this switch on (green) to have your iPhone automatically dial emergency services after its sensors detect a severe vehicular crash. The iPhone sounds an alarm and starts a countdown before placing the call, so you can turn it off if you're okay.

Exposure Notifications

On the Exposure Notifications screen, choose settings for warning you of possible exposure to health hazards, such as viruses. Set the Monthly Update switch on (green) to receive a monthly notification with updates. Set the Availability Alerts switch on (green) to receive a notification if exposure notifications are available in your region.

Battery

On the Battery screen (see Figure 5-11), you can configure settings for the battery and also see how it is performing:

- **Battery Percentage:** Set this switch on (green) to make the battery icon show the charge percentage as well as the approximate level.
- **Low Power Mode:** When the battery is getting critically low, set this switch on (green) to make the iPhone cut down on power consumption by reducing or disabling nonessential tasks and visual effects.
- **Insights and Suggestions:** If this box appears, it contains suggestions for cutting down on power usage — for example, by enabling the Auto-Brightness feature.
- **Usage Histograms and Data:** The statistics and histograms show the last charge level, the battery activity over the last 24 hours (see the left screen in Figure 5-11) or last 10 days (see the right screen in Figure 5-11), and the average times the screen was on and off. You can also see your activity and battery usage by app.

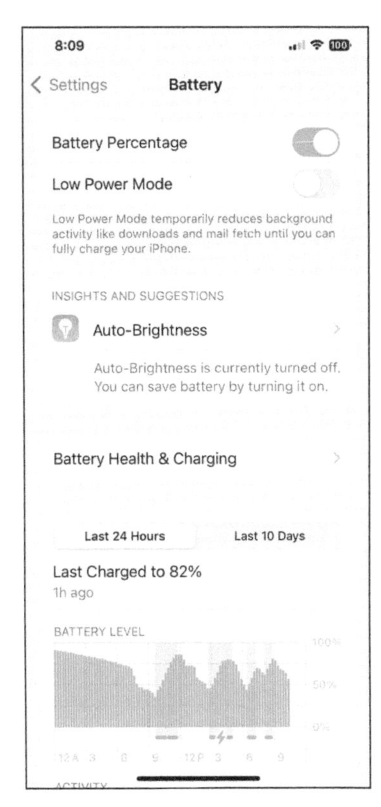

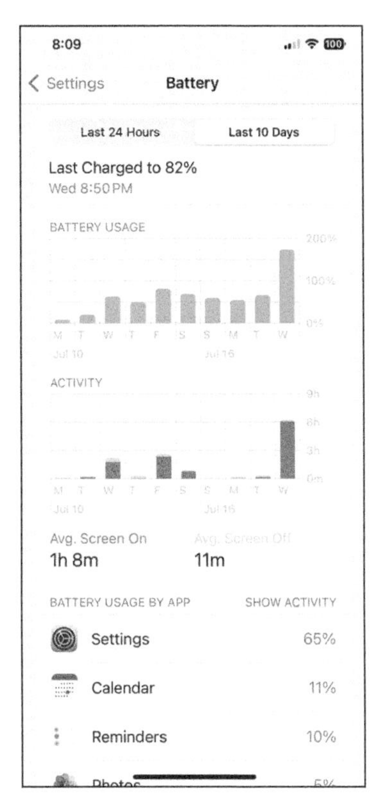

FIGURE 5-11: The Battery screen lets you see battery usage for the last 24 hours (left) or the last 10 days (right).

Tap the Battery Health & Charging button to display the Battery Health & Charging screen, which contains the following controls and information:

- **Maximum Capacity:** This readout shows how the battery's maximum capacity compares to the capacity when it was new — for example, 88%.
- **Peak Performance Capability:** This section displays the bland message "Built-in dynamic software and hardware systems will help counter performance impacts that may be noticed as your iPhone battery chemically ages." This means that the iPhone and iOS will try to coax the best performance out of the battery as it declines with age. (I can sympathize.)
- **Optimized Battery Charging:** Set this switch on (green) to have iOS study your charging routine and charge the battery fully only when you are likely to need it fully charged. At other times, iOS charges the battery to 80% and then stops charging so as to reduce wear on the battery.

TIP

If you need the battery charged as fully as possible all the time, set the Optimized Battery Charging switch and the Clean Energy Charging switch off (white). This may shorten the battery's lifetime and increase your carbon footprint.

- **Clean Energy Charging:** Set this switch on (green) to have iOS try to reduce your carbon footprint by changing when lower carbon emission electricity is available. In the Clean Energy Charging Helps Reduce Carbon Footprint dialog, suppress your guilt, and then tap either the Turn Off Until Tomorrow button or the Turn Off button.

Privacy & Security

The Privacy & Security screen enables you to choose settings for Location Services, tracking, individual apps and features used by other apps, and more.

Location Services

Location Services tracks your iPhone's location closely using GPS, cellular towers, Wi-Fi networks, and even Bluetooth connections. To control which apps and features can use Location Services, tap the Location Services button on the Privacy & Security screen. On the Location Services screen, work with the following controls:

- **Location Services:** Set this switch off (white) if you want to disable Location Services for everything except Find My iPhone. Normally, you'd keep Location Services on but allow only essential apps to use it.
- **Location Alerts:** Tap this button to display the Location Alerts screen. Here, set the Show Map in Location Alerts switch on (green) if you want to include the map in location alerts. Including the map is usually helpful.
- **Share My Location:** Tap this button to display the Find My screen. Tap Find My iPhone to display the Find My iPhone screen, and then set the Find My iPhone switch and the Find My network switch on (green). It's a good idea to set the Send Last Location switch on (green) to have your iPhone tell Apple its location when its battery is critically low.

  Back on the Find My screen, make sure the Share My Location switch is on (green) and that My Location button shows This Device if you want your iPhone to give your location (rather than having, say, your iPad provide your location).

  If the Family box appears, look at the list of family members. Each family member who appears in regular font can see your location; to cut them off, tap their button, and then tap Stop Sharing My Location. Your location is hidden from each family member who appears in gray font with the readout

Cannot see your location; to change this, tap their button, and then tap Share My Location.

» **Apps and Features:** The long list shows the apps and features that are using Location Services. To configure what an app or feature can do with Location Services, tap its button, and then choose settings on the screen that appears. For most apps, the Allow Location Access box appears, enabling you to choose among settings such as Never, Ask Next Time Or When I Share, While Using the App, or Always; in general, While Using the App is the best choice. If you allow the app or feature to use Location Services, set the Precise Location switch on (green) or off (white), as needed. Apps such as Maps need your precise location to be useful, but the precise location is overkill for any app or feature that just needs to confirm you're in a particular country.

TIP

At the very bottom of the Apps and Features list is System Services. Tap this button to display the System Services screen, on which you can enable or disable Location Services for a wide range of services. A *service* is a built-in function that performs actions in the background. For example, the Networking & Wireless service establishes, manages, and ends network and wireless connections.

Tracking

Tap the Tracking button on the Privacy & Security screen to display the Tracking screen. Here, you can choose whether apps can track your activities:

» **Allow Apps to Request to Track:** Set this switch on (green) *only* if you want to allow apps to ask to track your activities across other companies' apps and websites. Apps want to track you to learn more about you and build a profile of you. If you let apps request to track you, iOS displays a dialog (see Figure 5-12) for each request. Tap the Ask App Not to Track button or the Allow button, as appropriate.

» **List of apps that have requested to track:** In this list, set each switch on (green) or off (white) to control whether the app can track you.

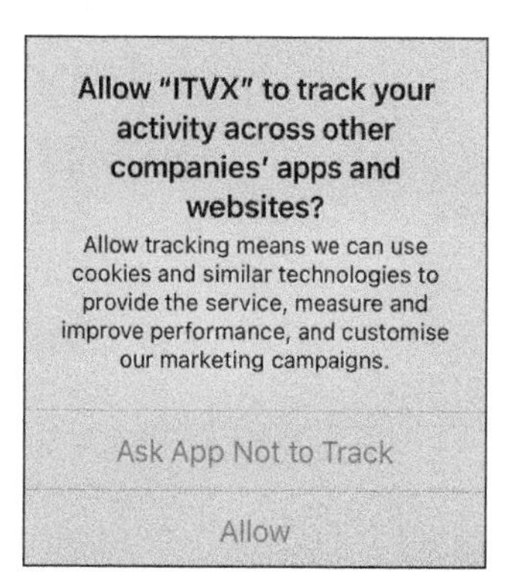

FIGURE 5-12: If you allow apps to ask to track you, choose Ask App Not to Track or Allow in dialogs such as this.

Privacy settings for individual apps and features

To choose privacy settings for an individual app or feature, tap its button in the long list. On the screen that appears, showing which apps and features have requested access to this app or feature, set each switch on (green) or off (white), as needed.

Safety Check

From the Safety Check screen, you can take two actions to increase your safety:

- » **Emergency Reset:** Tap this button, and then tap the Start Emergency Reset button to run through a process designed to protect your information from other people and from apps, to change your Apple ID password, and to review your emergency contacts to make sure they're the people you need now.
- » **Manage Sharing & Access:** Tap this button, and then tap the Continue button to go through a process of reviewing the people with whom you're sharing data, the apps that have access to your information (such as your calendar and contacts), and checking your account security.

Analytics & Improvements, Apple Advertising, and App Privacy Report

Toward the bottom of the Privacy & Security screen, you'll find these controls:

- » **Analytics & Improvements:** Tap this button to display the Analytics & Improvements screen. Here, you can set the Share iPhone Analytics switch, the Share iCloud Analytics switch, and the Share with App Developers switch on (green) or off (white) to specify what data you want to share anonymously with Apple and developers. Tap the Analytics Data button to see listings of the analytics data your iPhone has shared.
- » **Apple Advertising:** Tap this button to display the Apple Advertising screen. Here, you can tap the View Ad Targeting Information to display the information Apple's ads use about you. You can also set the Personalized Ads switch on (green) or off (white) to control whether Apple gives you personalized ads or just the regular old ads they give everyone. Either way, you still get the same quantity of ads.

- **App Privacy Report:** Tap this button to display the App Privacy Report screen, where you can enable and enable the App Privacy Report feature, which shows how often apps have used the permission you've granted them. For example, if you allow an app to access the Camera feature, App Privacy Report lets you see how often that app has done so.

Lockdown Mode

iOS, iPadOS, and macOS include a feature called lockdown mode that enables you to crank up an extreme level of protection against hacking and cracking threats. Lockdown mode is designed to protect your iPhone, iPad, and Mac against targeted attacks using custom malware or state-sponsored spyware.

Lockdown mode is designed for the tiny minority of users who face these types of threats. You may need to use lockdown mode if you're a human-rights activist, a crusading journalist, or a secret agent. If you're not exposed to such threats, lockdown mode is likely to be overkill.

To keep your devices safe, lockdown mode must limit some of the functionality that Apple users enjoy, such as the following:

- The Messages app blocks incoming message attachments, including link previews.
- FaceTime blocks calls and invitations from people outside your Contacts list.
- Safari's just-in-time compiler for the JavaScript programming language is disabled.
- The Photos app hides shared photo albums and blocks shared album invitations.
- iOS, iPadOS, and macOS disable wired connections to external devices or accessories while the device's screen is locked.
- The operating systems prevent you from configuring various sensitive settings, such as installing configuration profiles.

To put your iPhone into lockdown mode, tap Lockdown Mode at the bottom of the Privacy & Security screen, tap Turn On Lockdown Mode, and follow the prompts.

IN THIS CHAPTER

» Browsing for cool apps

» Searching for specific apps

» Getting apps onto your iPhone

» Deleting and organizing iPhone apps

» Reading books and news on your iPhone

Chapter 6
Finding and Managing Apps

One of the best things about the iPhone is that you can download and install apps created by third parties, which is to say not created by Apple (the first party) or you (the second party). At the time of this writing, around 2 million apps were available, of which about 400,000 are games. Some apps are free, other apps cost money; some apps are useful, other apps are lame; some apps are well behaved, other apps quit unexpectedly or worse. You'll never be short of choices for apps, but you may need to perform in-depth research to find good apps that do exactly what you want.

Where and How Do You Get Apps?

Apple restricts the iPhone to installing apps from the App Store, an online store that Apple curates and runs. App developers submit their apps to Apple for testing and approval, and Apple adds only tested and approved apps to the App Store. In theory, this means that all the apps should run well, without leaking memory and without crashing, and shouldn't take any user-unfriendly actions.

To access the App Store, you use the App Store app on your iPhone. The App Store app is part of iOS and normally appears on the first Home screen page by default.

TIP

If you don't find the App Store icon on the first Home screen page, look on each of the other pages, and then look in App Library. If you don't find the App Store app at all, the reason is most likely that someone else is managing your iPhone and has applied restrictions to it either via the Screen Time feature (as a family member might) or via a mobile-device management policy (as a company or organization might). Restrictions may hide the App Store app or may allow you to run the App Store app but not install apps; they might also prevent you from deleting installed apps.

The most straightforward way to install an app on your iPhone is installing it directly using the App Store app. But Apple also lets you set your iPhone to automatically install apps that you've installed on your other iOS or iPadOS devices that use the same Apple ID. So if you install an app on your iPad, your iPhone then installs it automatically, and vice versa.

To set up automatic downloads, choose Settings ➪ App Store, go to the Automatic Downloads section, and set both the App Downloads switch and the App Updates switch to on (green).

WARNING

Those switches apply only to Wi-Fi; to enable automatic downloads via cellular data connection too, go to the Cellular Data section of the App Store screen and set the Automatic Downloads switch on (green). Now tap App Downloads (also in the Cellular Data section) and choose Always Allow, Ask If Over 200MB, or Always Ask. If the controls in the Cellular Data section are disabled, wander up the App Store screen to the Allow App Store to Access section and set the Cellular Data switch on (green).

Finding, Installing, and Updating Apps

Finding apps for your iPhone is easy. To browse, search, download, and install apps, the only requirement is that you have an internet connection of some sort — Wi-Fi or cellular data.

To get started, tap the App Store icon on your iPhone's Home screen. After you launch the App Store, you see five icons at the bottom of the screen, representing five ways to interact with the store, as shown in Figure 6-1.

FIGURE 6-1: The icons across the bottom represent the five ways of interacting with the App Store.

Browsing for apps

Four of the five icons at the bottom of the screen offer ways you can browse or search the virtual shelves of the App Store. The fifth icon represents Apple's Arcade subscription gaming service; see the section "Visiting Apple's Arcade," later in this chapter.

Tap the Today icon and you'll find a variety of curated sections such as Meet the Developer and The Daily List. Scroll down to discover additional sections that may include Trending, Featured App, App and Game of the Day, and many others (sections change daily).

The Games, Apps, and Search icons offer different ways to view the App Store's offerings. The Games and Apps sections work much the same as the Today section, but their wares are limited to games and apps, respectively.

TIP

Each page displays dozens upon dozens of apps, but you see only a handful at a time on the screen. Remember to flick up and down and left and right to see all the others.

Know exactly what you're looking for? Tap the Search icon and type a word or phrase. Or tap the Apps icon and scroll down to browse by category. Also found in the Apps department are the helpful Top Free and Top Paid app lists, which display the most popular apps in the store today.

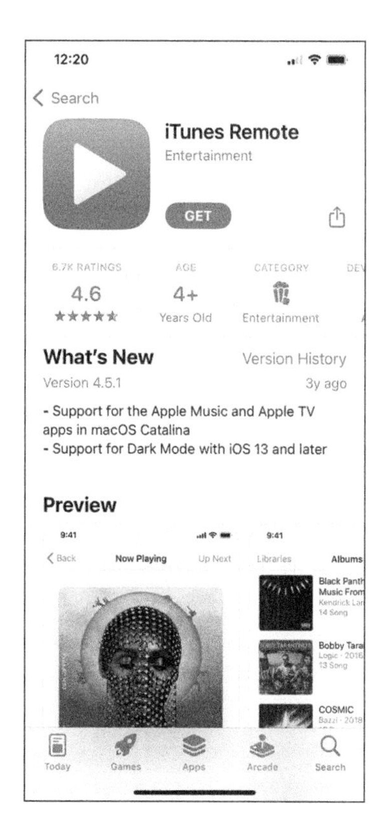

FIGURE 6-2:
iTunes Remote, a free app from Apple, lets you use your iPhone as a remote control for music on your Mac or PC.

Checking out an app's details

To find out more about any app on any page, tap the app. You see a details screen like the one shown in Figure 6-2.

TIP

Remember that the app description on this screen was written by the developer and may be somewhat biased.

Scroll down to see the Ratings & Reviews section (not shown in Figure 6-2) with star ratings and reviews for the app. Swipe left to see the next review; swipe right to see the previous one. Keep in mind that the reviews and ratings are from individual users rather than professional reviewers, so they may be biased, incomplete, or unhelpful.

Downloading an app

To download an app to your iPhone, tap the price button (for a paid app), Get button (for a free app), or cloud icon (for an app purchased previously or obtained for free) near the top of the app's details screen. If prompted to type your iTunes Store account password, do so. (You won't need to authenticate yourself for an app you've purchased or downloaded before.)

Once you are have authenticated yourself and have paid (if the app has a price), your iPhone downloads and installs the app. If you're still looking at the app's page in the App Store app, the Open button appears, and you can launch the app by tapping Open.

If you leave the App Store app and return to the Home screen, you can see the app being installed. The new app's icon is slightly dimmed and appears with a clocklike progress indicator, as shown on the left in Figure 6-3. Once the app is fully installed, its icon appears in all its glory, as shown on the right in Figure 6-3.

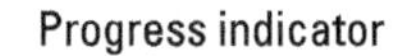

FIGURE 6-3: The progress indicator (left) tells you how much of the app has been downloaded; it disappears when the download is complete (right).

The app is now on your iPhone. Tap the app's icon to launch the app.

TIP

Once you've paid for an app, you can download it again and again if you need to, and you don't have to pay for it again, as long as the app is still available in the App Store. Better yet, if you've enabled Home Sharing or Family Sharing, your family members can download your apps (and you can download theirs) at no charge! See Chapter 5 for details on this pair of money-saving features for families.

Updating an app

Every so often (or, for some apps, far too often), the developer of an iPhone app releases an update. If one (or more) of these is waiting for you, a little number in a circle appears on the App Store icon on your Home screen. Tap the App Store icon, and then tap your account icon (which shows either your monogram or the picture you chose) in the top-right corner of the screen to open the Account screen.

Scroll down to the Available Updates section. If you see the Updated Recently section instead, none of the apps on your iPhone requires an update at this time. When an app does need updating, it will appear in the Available Updates section with an Update button to its right. Tap the Update button to update the app. If more than one app needs updating, you can update them all at once by tapping the Update All button.

TIP

Sections with your purchased apps, subscriptions, and recently updated apps also appear on the Account screen.

Finally, if you're using Family Sharing, you can see and search for apps purchased by family members from the Accounts screen as well.

Deleting an App

You may delete many of the Apple apps that came with your iPhone as well as any third-party app you no longer want or need.

Here's how to delete almost any app on your iPhone:

1. **Long-press any icon, and then tap Edit Home Screen on the pop-up menu to start the icons jiggling.**
2. **Tap the little – (minus sign) in the upper-left corner of the app you want to delete.**

   A dialog appears, informing you that deleting this app also deletes all its data.
3. **To remove the app from this device, tap Delete App.**

   If you instead tap Remove from Home Screen, the app will be removed from the Home screen and will appear only in App Library.

TIP

You can download any app you've purchased again, for free, from the App Store, provided that it's still available. So deleting it means only that it's gone from your device.

WARNING

If you see a warning that deleting the app also deletes any associated data, you may want to save the data before deleting the app. Different apps have different schemes for importing and exporting data. The important thing is that if you create documents with an app (notes, images, videos, and such), deleting the app may delete any files you've created with that app unless they're saved to iCloud or elsewhere. Forewarned is forearmed.

Unless, that is, you used the Offload Unused Apps feature in Settings ➪ App Store to automatically remove unused apps while saving their documents and data. Or, to offload individual apps manually, visit Settings ➪ General ➪ iPhone Storage and scroll down to the app, tap its name, and then tap its Offload App button. As long as the app remains available in the App Store, your documents and data will be restored when you reinstall the app.

Visit Settings ➪ General ➪ iPhone Storage regularly to monitor the available storage space on your iPhone.

If you delete an app and later change your mind and want it back on your iPhone, just open the App Store app, tap your account icon (which shows either your monogram or the picture you chose), and then tap Purchased. Then just tap the little cloud icon (shown in the margin) to reinstall the app on your iPhone.

Organizing Your Apps

You can have up to 19 Home screen pages, which gives you space for plenty of apps, although you'll likely want to devote some of that Home screen real estate to widgets as well. You can also store apps in the App Library instead of putting them on your Home screen pages.

Organizing your apps across Home screen pages

Once you've installed some apps, you'll likely need to organize them for easy access. This section shows you the moves to use.

To rearrange apps or create folders on your iPhone, long-press anywhere on any Home screen page, and choose Edit Home Screen, which makes the app and folder icons jiggle and dance and little "delete me" minus signs appear on the apps.

A folder icon works the same as an app icon when it comes to rearranging icons on-screen. Also, the app or folder you long-press to start the jiggling doesn't have to be the one you want to move — any app or folder will do. You can even long-press any open space on a Home screen page.

To move an app or folder after the jiggling starts, press it, drag it to its new location (other icons on the screen will politely move out of its way to make space for it), and release it.

If you move an icon onto another icon and pause for a second, iOS creates a folder. If you move an icon onto a folder and pause for a second, the folder opens so you can place the icon wherever you like in the folder.

Like the Home screen, folders can have more than one page. If you see two or more little white dots at the bottom of the folder — one for each additional page — swipe left to see the next page or swipe right to see the previous one.

To move an app to a different Home screen page after the jiggling starts, long-press the app and drag it all the way to the left or right edge of the screen. The preceding Home screen page or next Home screen page, respectively, will appear. Keep dragging the app to the left or right edge of each successive Home screen page until you reach the page you want. Then drop the app in its new location on that screen. If the screen is full already, the last icon on the page will be pushed to the next Home screen page. Be persistent — sometimes it takes a few tries to make the screens switch.

All these techniques work with apps on the dock (Phone, Safari, Messages, and Music by default), as well as with apps on Home screen pages and folders.

You can also add widgets to any Home screen page. To do so, long-press an app or folder and choose Edit Home Screen, then tap the + in the upper-left corner of the screen to see the Widget gallery overlay. Tap an item to display its details, and then swipe left or right to see the available sizes and layouts. When you've chosen a size and layout, tap the +Add Widget button to add the widget to the current Home screen page. The icons are still jiggling at this point, so you can move the widget to a different location or a different Home screen page by dragging it to where you want it.

After you have a few screens full of apps and folders, remembering where you put a particular app can be difficult. When that happens, three features can come to your rescue:

- **Search:** Search can quickly find and launch an app no matter which Home screen page it's on or which folder it's in.
- **Siri:** Just ask Siri to open the app by name.
- **App Switcher:** To switch quickly among your running apps, activate App Switcher by swiping upward from the bottom of the screen (Face ID) or double-pressing the Home button (Touch ID) to quickly switch between recently used apps.

If you're hazy on any of these three features, turn back to Chapter 2.

Managing apps with App Library

If you install lots of apps, managing and finding them may become difficult. To help, iOS provides App Library, which automatically organizes your apps into categories, as shown on the left in Figure 6-4. Tapping the search field at the top displays an alphabetical list of all apps on the device, as shown on the right in Figure 6-4.

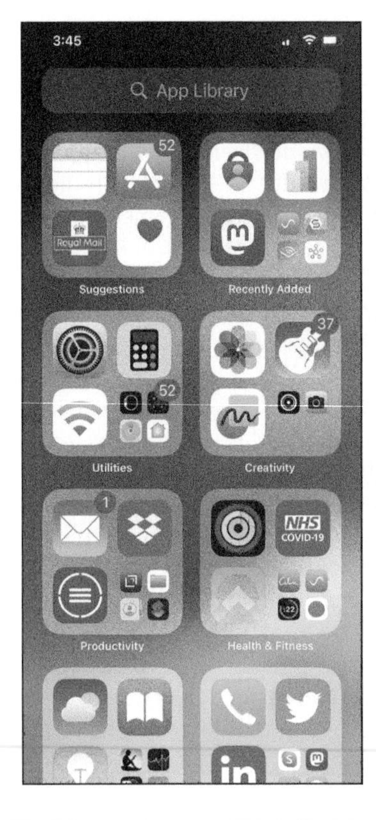

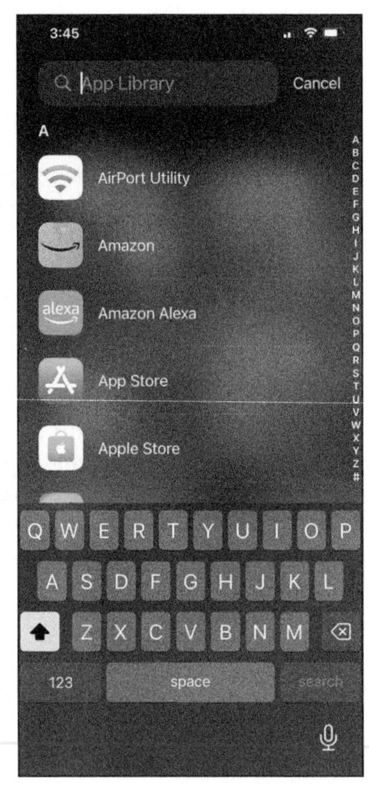

FIGURE 6-4: App Library organizes your apps into categories (left). You can also view them alphabetically (right).

To enable automatic organization, tap Settings ➪ Home Screen & App Library, and then tap either Add to Home Screen or App Library Only. And if you want apps in App Library to display notification badges like the little red 52-in-a-circle on the App Store icon in Figure 6-4, left, go to the Notification Badges section and set the Show in App Library switch on (green).

App Library appears after the last Home screen page. To display App Library, display the Home screen, and then swipe from right to left until App Library appears.

With App Library on screen, you can do the following:

- Tap any full-sized app icon to open it.
- Long-press any full-sized app icon to open its shortcut menu or delete the app.
- Long-press and drag any full-sized app icon to move it to either a different category or a Home screen page.
- Tap any group of miniature app icons to see all apps in that category.
- Tap the search field to search for an app by name.
- Swipe down or tap the search field to see an alphabetical list of your apps.

Understanding App Clips

An *app clip* is a small, self-contained part of an app that your iPhone can download and run without installing the entire app. An app clip fulfils a particular task, such as paying for a scooter rental or a food order.

Apple introduced app clips in iOS 14 in 2020, but they haven't really taken off, so they're still relatively rare in the wild. If you do run into an app clip, you invoke it by tapping a button in Mail, Messages, Safari, or another app built to support app clips; pointing your iPhone's camera at a QR code; or tapping an NFC tag with your iPhone. A banner appears describing the app clip and what it does. Tap the banner, and an overlay lets you choose either to use the app clip immediately or to visit the App Store to download the full version of the app.

If you choose to use the app clip right away (usually the best choice), it opens and behaves like any other app, allowing you to complete the task without visiting the App Store or downloading an app to your iPhone. When you're done, exit the app clip as you exit any app: Open App Switcher by swiping upward from the bottom of the screen or by pressing the Home button, then swipe the app clip up off the top of the screen. The app clip will remain in App Switcher unless you specifically close it in this way.

App clips can be a convenient way to perform a task that would otherwise require a visit to the App Store. So keep an eye out for the app clip logo (shown in Figure 6-5), identifying NFC tags, QR codes, and web links.

FIGURE 6-5: If you see this, an app clip is available by NFC or scan (left), or scan only (right).

Visiting Apple's Arcade

Apple's describes Arcade as "the world's first game subscription service for mobile, desktop, and the living room." With new, original games from legendary game creators Hironobu, Sakaguchi, Ken Wong, and Will Wright, Arcade offers benefits that include the following:

- Ability to play offline anywhere and anytime
- No ads
- No in-app purchases
- Access for up to six family members for $4.99/month
- Ability to switch from iPhone to iPad, to Mac, to AppleTV

If you're a gamer, visit `www.apple.com/apple-arcade/` for videos, game demos, and a link to your free 30-day trial.

Arcade is included in Apple One, Apple's bundle that includes some or all of its subscription services — Apple Music, Apple TV+, Apple Arcade, Apple News+, Apple Fitness+, and iCloud — at a lower price than you'd pay for them individually. For more info or a trial subscription visit `www.apple.com/apple-one`.

Enjoying Books, Newspapers, and Magazines

If you enjoy reading, your iPhone can be a great companion, as it enables you to carry any amount of books, newspapers, and magazines with you. Quite apart from being able to read anywhere, even in the dark, you can benefit from features such as changing fonts and type sizes, easy searching, and built-in dictionary lookup.

Your iPhone comes with Apple's Books app and News app installed. You can install other book and news apps from the App Store.

Books app

To start reading electronic books on your iPhone, tap the Books app on your Home screen. The app includes access to Apple's Book Store, which looks and works almost exactly like the App Store, allowing you to browse and shop for books 24 hours a day.

Apple Books and the Book Store aren't the only game in town. Check out other e-book options such as the Kindle, Nook, Bluefire, and Stanza apps, which many users prefer. Or try the Libby, Overdrive, and cloudLibrary apps, which let you borrow books from local libraries.

You buy Apple Books by tapping the Book Store button at the bottom of the Books app's main screen.

Apple Books purchased in this bookstore can be bought or read *only* on your Apple devices — your iPhones, iPads, or Macs running OS X 10.9 Mavericks or later. Sorry, Windows users, but there is still no Windows equivalent.

Newspaper and magazine apps

You can follow two paths to subscribe to or read a single issue of a newspaper or magazine. The first includes several fine publishing apps worth checking out, including USA TODAY, The Wall Street Journal, The New York Times, The Washington Post, Thomson Reuters News Pro, BBC News, and Popular Mechanics. You might also want to check out the free Zinio app, which offers publications including *Rolling Stone, The Economist, Macworld, PC Magazine, Car and Driver, National Geographic, Spin, Business Week,* and *Sporting News.* You can buy single issues of a magazine or subscribe, and sample and share some articles without a subscription.

You have to pay handsomely or subscribe to some of these newspapers and magazines, and most of them contain ads.

News app

The second path to periodicals is Apple's News app, a single app that gathers news articles, images, and videos you might be interested in and displays them in a visually appealing fashion. Participating publishers include ESPN, *The New York Times,* Hearst, Time, Inc., CNN, Condé Nast, and Bloomberg.

The first time you launch the app, you select topics that interest you. Then News creates a customized real-time newsfeed with stories it expects you to be interested in. The more you read, and the more you tell it which types of stories you want to see and which you don't, the better its suggestions should become.

In 2019, Apple introduced Apple News+, a subscription service with unlimited access to hundreds of magazines for $9.99 a month. Tap the News+ icon at the bottom of the screen to see a catalog of included publications or begin a free 30-day trial.

As mentioned earlier in the chapter, News+ is included in Apple One, Apple's bundle of subscription services. If you use subscription services heavily, Apple One might save you money.

2 Communicating and Organizing

IN THIS PART . . .

Enjoy voice calls with Phone and audio and video calls with FaceTime.

Browse the web with Safari.

Make the most of Messages and Notes.

Set up your email accounts and communicate via Mail.

Track your events with Calendar and your commitments with Reminders.

Find yourself and your way with Maps and Compass, and use other helpful tools.

IN THIS CHAPTER

» Making a call

» Using visual voicemail

» Receiving a call

» Facing up to FaceTime

Chapter 7
Making Phone and FaceTime Calls

Your iPhone enables you to stay in touch via voice anywhere you can get a cellular connection or a Wi-Fi connection. You can use the Phone app to make either person-to-person calls or conference calls involving multiple people. You can use the FaceTime app to make either video calls or audio calls with either a single other person or multiple people. If you're undecided about picking up a call, the Live Voicemail feature lets you see a transcription of the message even as the caller leaves it; and Visual Voicemail helps you to triage those calls you don't pick up. You learn how to do all this — and more — in this chapter.

Making a Call

To make a call, launch the Phone app by tapping its icon on the Home screen; the icon will be on the dock unless you've moved it. The Phone app opens, and you can make calls by tapping any of the five tabs at the bottom of the screen: Favorites, Recents, Contacts, Keypad, or Voicemail, from left to right. You can also place calls by voice, using Siri.

TIP

On the iPhone 14 Pro and iPhone 15 Pro models, if you need to summon emergency services but there's no signal, tap Emergency Text via Satellite in the bottom-right corner of the call screen. On the screen that appears, tap Report Emergency, and then follow the prompts to specify the predicament you're in and request assistance. This service is available for two years after you activate your iPhone.

Making calls from the Contacts screen

The Contacts tab in the Phone app gives you direct access to the contacts you've added through iCloud and your other contacts-capable accounts. You can quickly make calls to your contacts like this:

1. **In the Phone app, tap Contacts to display the Contacts tab.**
2. **Locate the contact you want to call by scrolling down, tapping the appropriate letter on the right side of the screen, or starting to type the contact's name in the Search field at the top.**

   To move a long way through the list, you can also flick your finger up the screen to start momentum scrolling, and then tap the screen to stop the scrolling.
3. **Tap the contact to display their contact information.**
4. **Tap the phone number to call (see the left screen in Figure 7-1).**

   The Phone app places the call. If the contact answers, greet them civilly. The readout at the top of the screen shows the duration of the call.
5. **When it's time to end the call, tap the End button.**

Stripped down to those five steps, making a call could hardly be easier. But while we're talking about contacts, let's look quickly at how you can navigate the Contacts screen's lists and how you can create contact posters for your contacts.

TIP

Your own iPhone phone number appears at the top of the Contacts list inside the My Card listing.

Navigating the Contacts screen's lists

The Contacts app and the Contacts screen in the Phone app enable you to assign your contacts to different lists and display either some or all of those lists. For example, you might create a Family list, a Friends list, a Work list, and so on.

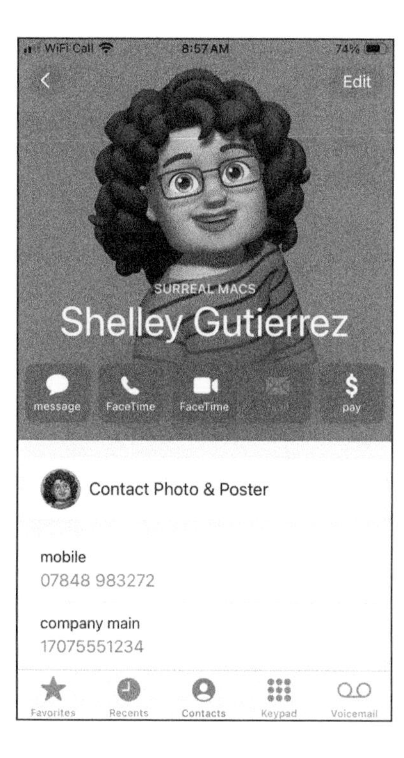

FIGURE 7-1: In the Phone app, tap the number to call (left). Tap the End button to end the call (right).

The name at the top of the screen shows which contacts you're working with. If the screen is called Contacts, you're viewing the All Contacts list. If the screen shows another name, you're working with that list.

To change lists, display the Lists screen by tapping the Lists button at the top of the Contacts screen, and then tap the list you want to use. From the Lists screen, you can start a new list by tapping the Add List button in the upper-right corner, choosing the account in which to store the list (if you're using multiple contacts accounts), and then typing the list name. To add contacts to the list, tap the list, and then tap Add Contacts on the list's screen.

From the Lists screen, you can delete a list by swiping left on its button, and then tapping the delete icon (trashcan) that appears.

Creating contact posters

NEW

In iOS 17, the Contacts app enables you to create *contact posters*, custom screens that appear when you display a contact card or are communicating with the contact via a phone call or a FaceTime Audio call. Contact posters not only provide visual interest but also help you triage incoming calls more quickly. If you frequently switch among calls, you may also find contact posters help you keep tabs on who you're talking to right now.

To create a contact poster, follow these steps:

1. **In the Contacts app or on the Contacts screen in the Phone app, tap the contact to display the contact's screen.**
2. **Tap the Contact Photo & Poster button to display the Contact Photo & Poster screen.**

   The Contact Photo & Poster screen appears for a contact to whom you haven't yet assigned a photo (or a memoji or a monogram) or a contact poster. If you've already assigned the contact a photo (or a memoji or a monogram) or a contact poster, you'll see a different screen showing that image or the image and the contact poster. Tap the Customize button to change either the photo or the poster.
3. **In the Choose Your Poster area, assign a photo, a memoji, or a monogram to the contact.**

   You can tap the Camera button to take a photo, tap the Photos button to use an existing photo from the Photos app, tap the Memoji button to create a new memoji or use an existing one, or tap the Monogram button to create a new monogram (the color circle containing initials) or customize the existing one. This example uses a new memoji.
4. **On the Preview Poster screen (shown on the left in Figure 7-2), tap the Continue button.**
5. **If the contact record already contained a poster, you get a choice:**
   - *Tap the Update button to proceed with the poster shown in the preview.*
   - *Tap the Choose a Different Poster button and select a different poster.*
6. **On the Contact Photo screen, tap the Crop button to display the Move and Scale screen (shown on the right in Figure 7-2).**
7. **Drag the image as needed so that the circle contains the part you want to display. Pinch in or out to change the image size.**

   For a photo, swipe left to try the different color filters available.
8. **Tap the Choose button.**
9. **Tap the Done button when you've finalized the poster.**

   The Contacts app applies the photo and poster to the contact. Now, when you call the contact or the contact calls you, the Phone app displays the contact poster.

FIGURE 7-2: On the Preview Poster screen (left), tap the Continue button. On the Move and Scale screen (right), position the image in the circle and scale it as needed.

Setting up favorites (and calling them)

If you call a particular contact frequently, add them to your Favorites list. You can then access them quickly from the Favorites screen in the Phone app.

TIP

You can set up as many favorites as you need for a person. So, for example, you might create separate favorite listings for your spouse's office phone number, cell number, and FaceTime entry.

To designate a contact a favorite, follow these steps:

1. **Tap the contact's record to open it.**
2. **Scroll down to the bottom of the record.**
3. **Tap the Add to Favorites button to display the Add to Favorites pop-up menu (see Figure 7-3, left).**

   The Add to Favorites button doesn't appear if the contact record contains no means of contact — a phone number, an email address, or whatever.

4. **Tap Message, Call, or Mail to specify which type of favorite to create.**

   The entries of that type appear (see Figure 7-3, middle).

5. **Tap the address or phone number you want to make a favorite.**

   Contacts displays a favorite star to the right of the address or phone number, as you see for the iPhone phone number in Figure 7-3, right.

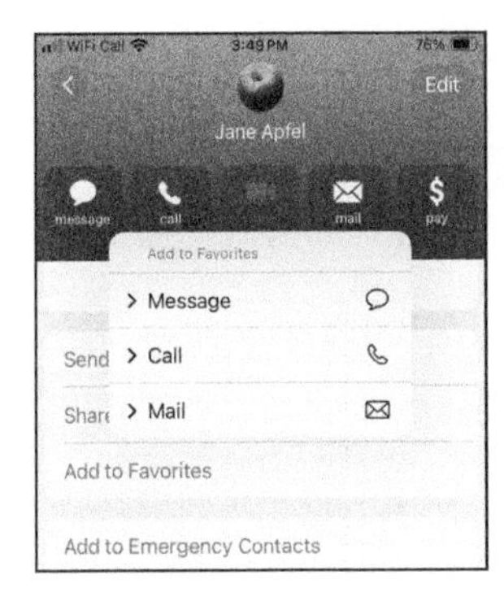

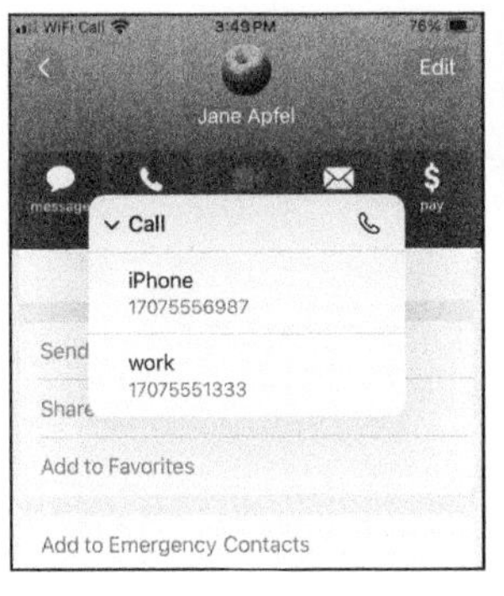

FIGURE 7-3: To create a favorite, tap the type, such as Call (left), and then tap the number or address (center). The favorite displays a star (right).

To display all your favorites, tap the Favorites tab at the bottom of the Phone app.

To remove favorite status from a contact's entry, swipe their button on the Favorites tab all the way to the left. You can also swipe it partway left, and then tap the Delete button.

TIP

To rearrange your favorites into your preferred order, tap the Edit button in the upper-right corner of the Favorites screen, and then drag a favorite up or down by its handle (three horizontal lines). When you're done, tap the Done button.

To call a favorite, tap the appropriate entry on the Favorites screen.

Making calls from the Recents list

Tap the Recents tab at the bottom of the Phone app to display the Recents list, which logs your recent incoming and outgoing calls. Tap the All tab at the top to display all calls, or tap the Missed tab to see only the calls you missed. On each tab, missed calls appear in red.

To see the details available for a call, tap the *i*-in-a-circle next to an item. From the screen that appears, you can tap the Create New Contact button to create a new contact with the available information, tap the Add to Existing Contact button to add this information to a contact you choose, or tap the Share Contact button to share the contact with other people.

To return a call, tap anywhere on the main part of the button (not the *i*-in-a-circle).

To remove an entry from the Recents list, swipe partway left on the entry, and then tap the delete icon (trashcan). Alternatively, swipe left all the way.

To clear the Recents list, tap the Edit button in the upper-left corner, tap Select, and then tap the Clear button.

TIP

You can dial a call by voice by asking Siri to place the call. For example, say, "Hey, Siri! Call Alice Smith on her mobile number for me."

Dialing with the keypad

To dial a number manually, or at least digitally, tap the Keypad tab to display the keypad (see Figure 7-4), press the keys for the number, and then tap the green call button.

FIGURE 7-4:
Dialing with the keypad.

If the Phone app recognizes the number you're dialing as being one of your contacts, it displays the contact's name at the top of the Keypad screen. If it doesn't recognize the number, it displays the Add Number button, which you can tap to add the number to your contacts.

TIP

To repeat the last number dialed, long-press the green call button.

Accessing voicemail visually

Apple's Visual Voicemail feature enables you to tackle your voicemail messages in whichever order you prefer rather than having to listen to them in the order received. A red badge on the Voicemail tab at the bottom of the Phone screen shows how many messages are waiting for you. Tap the Voicemail tab to display the list of messages. A blue dot to the left of a message indicates that you haven't yet listened to the message. Tap the message you want to listen to.

USING YOUR iPHONE AROUND THE WORLD

If you travel, you'll likely want to use your iPhone wherever you go. Keep these three points in mind:

- For top speeds, you'll want a 5G-capable iPhone. That means an iPhone 12 or later model or a second-generation or later iPhone SE.
- If your iPhone takes a physical SIM card, you can switch SIM cards to connect to a different carrier's network if necessary. However, if you do need to connect to a different carrier's network, you may be better off getting an eSIM, a virtual SIM card.
- Most current iPhones for the US and Canadian markets use eSIMs rather than physical SIM cards. If your iPhone is one of these models, eSIMs are your only way to connect to a different carrier's network.
- To connect your iPhone to a different carrier's network without adding a physical SIM card or eSIM, you'll need to get your carrier to turn on international roaming for your account. Before setting this up, make sure you know how many arms and legs it will cost.
- If you're calling the US while overseas, turn on the Dial Assist feature to simplify international dialing. Choose Settings ➪ Phone, and then set the Dial Assist switch on (green).

REMEMBER

Visual Voicemail is available only from certain carriers. If your carrier doesn't support Visual Voicemail, you'll need to access your voicemail the old-school way.

To record a custom greeting for your voicemail, tap the Voicemail tab, tap the Greeting button in the upper-left corner, and then tap Custom.

To return a call, tap the call back icon (blue circle with a receiver).

To delete a voicemail message, tap the delete icon (red circle with a trashcan).

Receiving a Call

When someone calls you, you can answer the call or decline it as easily as falling off a log.

NEW

To help you decide whether to pick up the call, iOS 17 provides the Live Voicemail feature. If your carrier supports this feature, your iPhone displays either the voicemail icon in the dynamic island (on iPhone models with dynamic island) or the phone icon in the status bar (on iPhone models without Dynamic Island) when someone is leaving you a voicemail message. Tap this icon to display a screen showing a transcription of the message. There's a delay of a couple of seconds, but the feature is fast enough to give you a fair chance of catching all but the tersest callers. Tap the accept icon (green circle containing a white handset) if you want to pick up the call while the caller is leaving their message.

Accepting the call

To accept a call, you have four options:

- Tap the accept icon (green circle containing a white handset).
- If the phone is locked, slide the Slide to Answer slider to the right.
- If you're wearing stereo earbuds or EarPods, tap the microphone icon or button. Microphone adapters for standard headsets may also work.
- If you wear a wireless Bluetooth headset or use a car speakerphone, push the Answer button on your headset or speakerphone. With Apple's AirPods, double-tap one of the earbuds; with AirPods Pro, press the indent on the stem.

TIP

REMEMBER

If you're listening to music in the Music app or another audio app when a call comes in, the song pauses playing while you decide whether to take the call. If you do take the call, after the conversation ends, the music usually resumes from where it paused.

Announcing the caller

There are three main ways of telling who is calling you before you pick up. The first way is Caller ID. The second way is recognizing the ringtone you have assigned to a particular caller (see the section "Choosing a ringtone," later in this chapter).

The third way is to have Siri announce the caller's name. To enable this feature, choose Settings ⇨ Phone ⇨ Announce Calls, and then tap Always, Headphones & Car, or Headphones Only to tell Siri when to make the announcements. Choose the fourth option, Never, when you need to keep Siri quiet — for example, when you're in public.

Rejecting the call

If you don't want to pick up, reject the call in one of these ways:

- » Tap the red decline button, which appears inside a notification at the top the screen when your iPhone is unlocked (see in Figure 7-5).
- » Let the call ring until it goes to voicemail.
- » Press the side button twice in rapid succession.
- » Press and hold down the microphone button on a headset for a couple of seconds, and then let go. Two beeps let you know that the call was indeed rejected. On some Bluetooth earpieces, pressing and holding down other buttons does the trick.

FIGURE 7-5:
You can tap the red or green buttons to reject or answer a call.

Replying with a text message

When you can't take a call, consider replying with a text message. When the banner notification for the incoming call appears, swipe down on the notification to display the call screen, and then tap the Message icon. On the Respond With pop-up menu (see Figure 7-6), either tap one of the three canned messages or tap Custom and type a custom message.

TIP

To customize the three canned messages, choose Settings ⇨ Phone ⇨ Respond with Text, and then type your preferred messages on the Respond with Text screen.

When you tap any of these options, the caller will get your outgoing voicemail and receive whichever of the text options you chose to send.

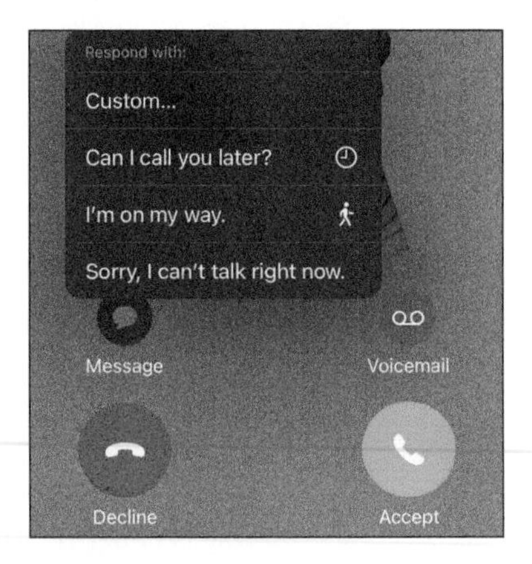

FIGURE 7-6:
When you can't take a call, respond with a text message.

REMEMBER

If the incoming call is from an unknown person and a phone number isn't shown, you won't have the option to reply with a text. Your iPhone is smart — but not smart enough to guess who the caller might be. If you tap Remind Me instead of Message, you can arrange to receive a reminder notification in one hour. Meanwhile, the caller hears your voicemail message. Or you can arrange a reminder when you leave your current location.

To turn off the ringer, set the Ring/Silent switch on the left side of your iPhone so that orange is visible. If you've forgotten to do this and get an incoming call at an inopportune moment, squelch the ring and the vibration by pressing the side button or one of the volume buttons once. You'll still be able to answer the call (discreetly, please).

When you have enabled the driving focus in Settings (see Chapter 5), your iPhone sends an automatic text to any call that comes in.

Blocking callers

When you need to tune out calls from a particular caller, tap the Recents tab, and then tap the *i*-in-a-circle on a call from that caller. Scroll down to the bottom the screen for that caller, tap the Block This Caller button, and then tap the Block Contact button to add the caller to the Blocked Contacts list. Your iPhone then sends any future calls from the caller straight to voicemail.

To review your Blocked Contacts list, choose Settings ⇨ Phone ⇨ Blocked Contacts. On the Blocked Contacts screen, you can quickly unblock a contact by swiping partway left on the contact and then tapping Unblock, or by swiping all the way left, so that the Unblock button grows and engulfs the contact's entry.

Silencing unknown callers

When you want to hear only from those contacts you haven't blocked, turn on your iPhone's Silence Unknown Callers feature. Choose Settings ⇨ Phone ⇨ Silence Unknown Callers, and then set the Silence Unknown Callers switch on (green).

iOS then sends all calls from unknown numbers straight to voicemail. The calls still appear in the Recents list, so you needn't worry about missing a vital call.

Avoiding disturbances but permitting essential calls

When you don't want to be bothered by a phone call, enable the Do Not Disturb feature or one of the other Focus settings. To do so, open Control Center, tap the main part of the Focus button (not the icon), and then tap Do Not Disturb button or the button for the other focus you want.

You can configure Do Not Disturb or any other focus to allow calls from anybody you still need to hear from. To do so, choose Settings ➪ Focus, and then tap the focus you want to configure. On that focus's configuration screen, go to the Phone Calls box, tap the Allow Calls From pop-up menu, and then tap the group allowed to contact you — for example, Favorites. If you don't want to use a group, open the Allow Calls From pop-up menu, and then tap Allowed People Only. Next, tap the Add People button, tap to select the selection circle for each contact to allow, and then tap the Done button.

Still on the focus's configuration screen, set the Allow Repeated Calls switch on (green) if you want a second call from the same person within three minutes to break through the focus. The assumption is that repeated calling indicates something seriously amiss. Set the Allow Repeated Calls switch off (white) if your contacts suffer from impatient chromosomes.

You can also configure Emergency Bypass for individual contacts. Emergency Bypass sounds like critical surgery likely to carry a ferocious deductible, but on the iPhone it lets a contact's phone calls or instant messages bypass your Do Not Disturb and Focus settings. Tap the contact to display their details, and then tap the Edit button. Tap the Ringtone button to display the Ringtone screen, and then set the Emergency Bypass switch on (green). In the Ringtones section, select a distinctive ringtone for the content, and then tap the Done button. Next, tap the Text Tone button to display the Text Tone screen, and then lather, rinse, and repeat.

Choosing a ringtone

Your iPhone comes with more than 50 ringtones, so you have a good chance of finding one that you like. To set the ringtone, choose Settings ➪ Sounds & Haptics ➪ Ringtone, and then tap each ringtone in the Ringtones box and the Alert Tones box until you find one you like. Tap the Classic button at the bottom of the Ringtones box and the Alert Tones box to reach the classic tones that came with early iPhone models.

While on the Ringtone screen, tap the Haptics button, and then pick a haptic (vibration pattern) to accompany the ringtone.

Tap the Back button to return to the Sounds & Haptics screen. You can then choose tones and vibrations for other items in the Sounds and Haptic Patterns box — Text Tone, New Voicemail, New Mail, and so on.

ADDING RINGTONES

Even with all the ringtones your iPhone includes, you may want to add more. You can purchase and download ringtones wirelessly from your phone via the iTunes Store. From the Home screen, tap iTunes Store, and then tap Tones. (If you don't see Tones at first, tap the More button at the bottom right of the iTunes Store screen.) Alternatively, tap Settings ⇨ Sounds & Haptics ⇨ Ringtone ⇨ Tone Store. You can find ringtones by genre, through top charts (in a given musical category), and by browsing the selections featured in the store. You can use any ringtone as a text tone.

The ringtone and text tone you've just set are the defaults for all incoming calls. To distinguish individual callers, you can assign specific ringtones. On the Contacts screen, tap the contact's record to open it, and then tap the Edit button. Tap the Ringtone button, tap the ringtone you want, and then tap the Done button. Back in the contact record, you can tap Text Tone if you want to assign a custom text tone too. Tap the Done button when you've finished tweaking the contact.

While on a Call

While you're making a call, you can switch to other apps by using either App Switcher or the Home screen. You can also

- **Mute a call:** From the main call screen (shown on the right in Figure 7-1), tap Mute. Tap Mute again to unmute the sound.
- **Place a call on hold:** Long-press the Mute icon until it changes to a Hold icon. Tap the Hold icon to take the person off hold. You might put a caller on hold to answer another incoming call or to make a second call yourself.
- **Tap Add to display the Contacts list:** The Contacts list enables you to add someone to a conference call or simply look up information during a call.
- **Tap Keypad to display the keypad:** Use the keypad to navigate through an automated menu system. Tap the Hide Keyboard button to return to the main call screen.

» **Use the speakerphone:** Tap Speaker to listen to a call through the iPhone's internal speakers without having to hold the device up to your ear. If you've paired the iPhone with a Bluetooth device, the control is labeled Audio instead. Tap Audio and then tap Speaker (if you want the speakerphone), iPhone (if you want to hold up the phone to your ear), or the name of the Bluetooth device. A check mark appears next to your selection.

» **Make a conference call:** Read on.

Juggling calls

When you're already on the phone with someone and another call comes in, you can take the following actions (see the left screen in Figure 7-7):

» Tap the Send to Voicemail button to send the incoming call to voicemail.

» Tap the End & Accept button to end the current call and take the new call.

» Tap the Hold & Accept button to put the current call on hold and take the new call. You can then toggle between the calls by tapping either the Switch button or the call on hold at the top of the screen (see the right screen in Figure 7-7).

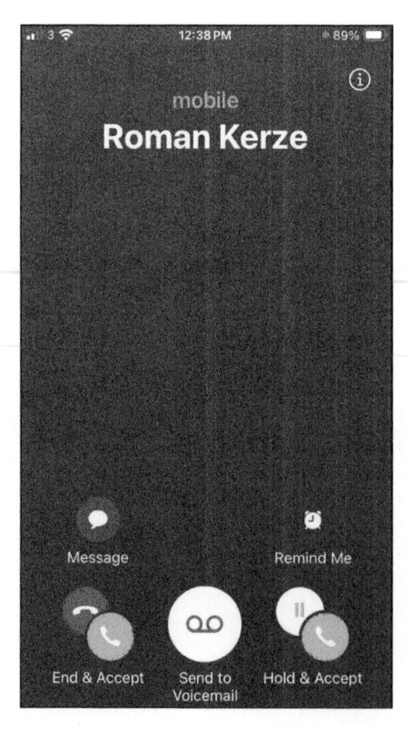

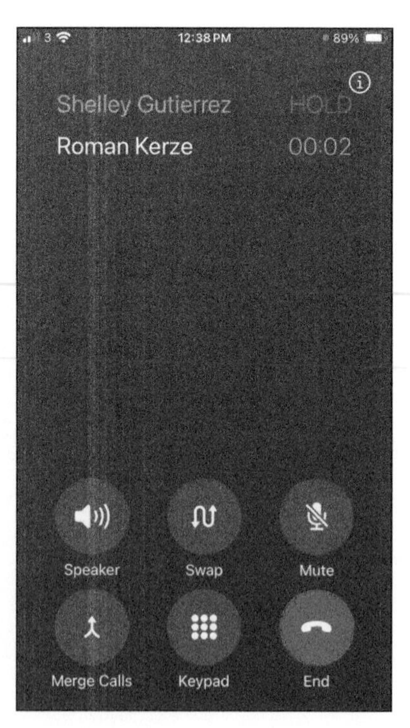

FIGURE 7-7: Handling two calls at once.

Making conference calls

From having one call active and one call on hold, you can tap the Merge Calls button to transition seamlessly to a conference call. Once your iPhone has merged the calls, you see both callers' names at the top of the screen (see Figure 7-8, left).

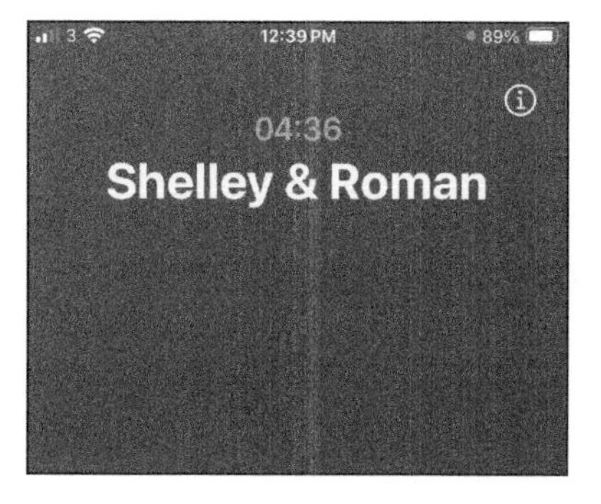

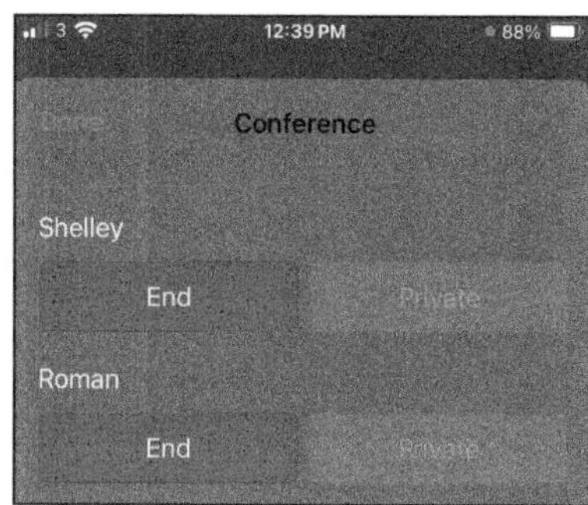

FIGURE 7-8: In a conference call, the callers' names appear at the top (left). Tap the *i*-in-a-circle to display the conference screen (right).

If you need to add another person to the call, tap the Add Call button, and then dial the contact, which puts the other callers on hold. Once you've gotten hold of the person, tap the Merge Calls button to reestablish the conference call, including the new person.

If you need to drop one of the callers, tap the *i*-in-a-circle to display the Conference screen (see Figure 7-8, right), and then tap the End button for that caller. Alternatively, tap the Private button to speak privately to that caller, which puts the other callers on hold. Tap the Merge Calls button when you're ready to return to the conference call.

TIP

The number of people you can add to a conference call depends on your carrier, not directly on your iPhone. Between five and ten people is typical.

Handing off calls

Apple's Handoff feature enables you to pass tasks between your iPhone, your iPad, and your Mac. This integration means that you can use your iPad or your Mac to make or answer a call that originated from or was received by your iPhone.

Each device must be signed into the same iCloud account, must be connected to the same Wi-Fi network, and must have Handoff enabled. To enable Handoff on the iPhone or iPad, choose Settings ➪ General ➪ AirPlay and Handoff, and then set the Handoff switch on (green). On the Mac, choose System Settings ➪ General ➪ AirDrop & Handoff, and then set the Allow Handoff Between This Mac and Your iCloud Devices switch on (blue).

When you're using your Mac or your iPad, you'll see the name, phone number, and profile picture of a person calling your iPhone. From the Mac or iPad, click (or swipe) the notification of the call to answer or reject the call.

To make a call from your Mac or iPad, click or tap a phone number in the Contacts, Calendar, or Safari app.

Making and Taking FaceTime Audio and Video Calls

FaceTime enables you to make audio-only or video-and-audio calls from your iPhone, iPad, or Mac. On the iPhone, using FaceTime is as easy as making a regular call — and FaceTime offers two major benefits quite apart from the video:

- FaceTime calls don't count against your call allowance, though they will count against your data plan allotment if you make these calls over a cellular connection.
- The audio quality on FaceTime calls, at least those over Wi-Fi, can be superior to a regular cellphone connection. But a robust 5G or other cellular connection typically works fine too.

FaceTime even lets you include users of non-Apple devices, such as Android and Windows, in calls.

Making a FaceTime call

Here's how to make a FaceTime call:

1. **Tap the FaceTime icon on the Home screen to launch the FaceTime app.**

   The FaceTime screen lists recent calls you made (see Figure 7-9, left). To make one of those calls again, tap its button and skip Steps 2–4.

2. **Tap the New FaceTime button to display the New FaceTime screen (see Figure 7-9, right).**
3. **Choose your victim by tapping one of the Suggested buttons; by typing a phone number or email address in the To field; or by tapping the circled +, and then adding contacts from the list that appears.**

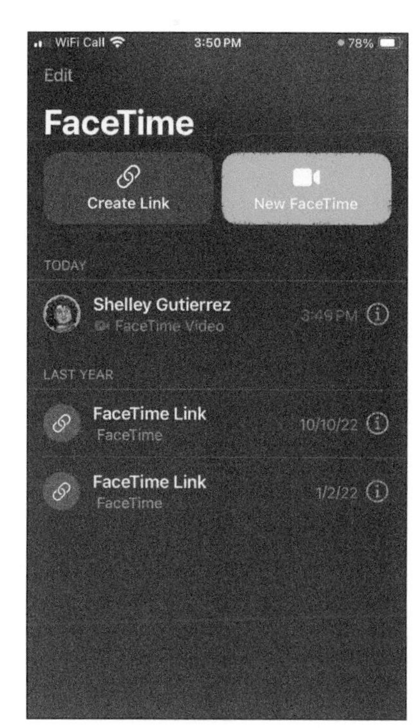

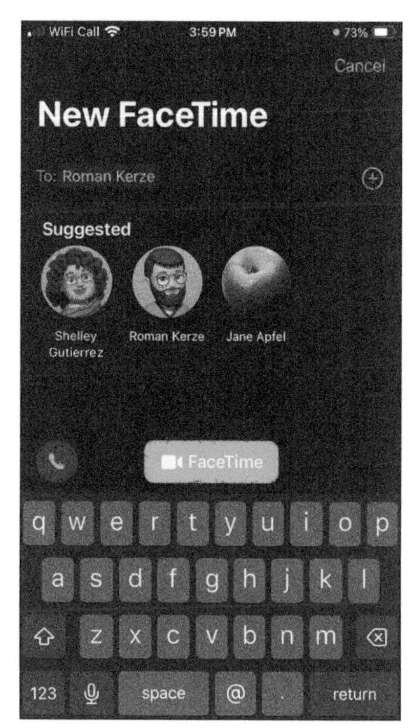

FIGURE 7-9: Choose whom to FaceTime and what kind of call to place.

4. **Tap the FaceTime button to start a FaceTime video call, or tap the telephone icon to make a FaceTime Audio call.**

   FaceTime places the call. The other person can accept the call by tapping the green button or decline it by tapping the red button. If the other person is using an iPhone and it's locked, they can slide a slider to accept the call. They can also decline and send you a text message or ask to be reminded to get in touch later.

To invite someone with a non-Apple device to FaceTime with you, tap the Create Link button and share that link with the person. When they try to join the call, you get to approve the connection.

TIP

Here are two other ways to place FaceTime calls. First, ask Siri: Just say "Hey, Siri, make a FaceTime call to Dad" or a similar command. Second, in your Contacts list, tap the FaceTime icon (video camera) or the FaceTime audio icon (telephone).

Receiving a FaceTime call

When you receive an incoming FaceTime request, tap the green button, and then tap the Join button; if your iPhone is locked, slide the Slide to Answer slider. You'll then see and hear the caller and be able to communicate with them. Your own video feed appears in a small picture-in-picture (PiP) window, enabling you to check for spinach in your teeth and (if you find some) get out of the frame, stat.

TIP

You can use FaceTime in portrait mode or landscape mode. You might find it easier to bring another person into a scene in landscape mode.

TIP

You can also use the front or rear cameras during a FaceTime call. To toggle between the front and main cameras, tap the thumbnail and then tap the switch cameras icon (still camera). To mute a FaceTime video call, tap the microphone icon with the slash running through it. The caller won't be able to hear you but can continue to see you.

If you have an iPhone X or later model, you can add cartoonish *memojis* or stickers. (Read about them in Chapter 9.) And you can add shapes, filters, arrows, and even text during a FaceTime call. Tap the choose effects icon (wavy star) to get going. Figure 7-10 shows how memojis solve the problem of the author trying to hold both ends of a FaceTime chat.

FIGURE 7-10: FaceTime offers a high-quality video connection even if you don't use memojis.

Although many FaceTime calls commence with a regular phone call, you can't go from a FaceTime video call to an audio-only call without hanging up and redialing.

From the FaceTime app, you can also designate FaceTime favorites (audio or video), and make FaceTime calls from the Recents screen of the Phone app.

To block all FaceTime calls, choose Settings ➪ FaceTime, and then make sure the FaceTime switch is off (white). While you're in FaceTime Settings, you can list one or more email addresses by which a caller can reach you for a video call, along with your iPhone's phone number.

If you want to momentarily check out another iPhone app while on a FaceTime call, press the Home button on a Touch ID iPhone model or swipe up on a Face ID model, and then tap the icon for the app you have in mind.

Through the picture-in-picture feature, you can continue to peek at the person you're talking to in a window you can drag around the screen while exploring other apps and functions.

If you like what you see, snap a photo of the FaceTime call in progress.

Making group FaceTime calls

FaceTime enables you to have group chats containing up to 32 people (including yourself). You can either start a call with multiple people or start small and then add people after establishing the call. The more people in the call, the smaller the boxes on the screen that show the people. However, grid view lets you see up to a half-dozen people on the screen at once in same-size tiles, which may make conversing feel more natural.

The easiest way to initiate a group FaceTime call is to start a group chat in the Messages app (see Chapter 9).

Recent improvements to FaceTime

Over the last several versions of iOS, Apple has made huge improvements to FaceTime. The following three are the improvements you'll likely benefit from most:

- » Portrait mode blurs the background behind a person so you can focus on that person, inspired by a similar feature in the Photos app (see Chapter 14).
- » Machine learning helps your iPhone isolate participants' voices while turning down the volume of background noises such as traffic, sirens, or barking dogs. FaceTime applies spatial positioning to the audio, so voices sound like they're coming from the directions where people are positioned on the video call.

- Through the SharePlay feature, you and your co-callers on FaceTime can share your screens, stream music, or watch TV or a movie together on a call. To share media, you first open a streaming app that supports SharePlay, such as the Music app or the TV app; choose the song, show, or movie to share; tap the share icon (an arrow escaping from a box) to display the Share sheet; and finally tap the SharePlay button and follow the prompts that appear. You can share content only with people who can access the same streams via their own subscriptions. During a call, you or others may have to tap Join SharePlay, depending on who initiated SharePlay. You also have the option to watch or listen to stuff during a FaceTime call without sharing it with other people.

IN THIS CHAPTER

» Surfing the web

» Opening and displaying web pages and tabs

» Making the most of with links, bookmarks, and History lists

» Separating your browsing into different profiles

» Securing Safari

Chapter 8 Going on a Mobile Safari

In this chapter, you learn to surf the web with Safari for iOS, the iPhone version of Apple's sleek and powerful web browser. Safari is (arguably) one of the best smartphone browsers and (unarguably) enables you to enjoy all the features of most websites. You can open multiple web pages at the same time, organizing them with tabs and tab groups; bookmark interesting web pages and share them among your devices that log into iCloud; and manage and use your browsing history. When discretion is the better part of valor, you can use private browsing to cover some of your tracks; use the Sign In with Apple feature to avoid giving your email address to sites you don't trust; use profiles to split your work browsing from your home browsing; and configure Safari to protect your privacy.

Starting a Safari into Cyberspace

Before you launch Safari, take a look at Figure 8-1. Safari on the iPhone includes only some of the browser controls typically found on a PC or Mac, and Apple has streamlined the tab bar (also called the *address bar*) and positioned it at the bottom of the display by default, within easy reach of your thumb. To free up space, the tab bar hides automatically when you scroll down, but you can make it reappear by scrolling up again.

The tab bar contains the smart search field, a combined address bar and search field in which you can type web addresses or search terms.

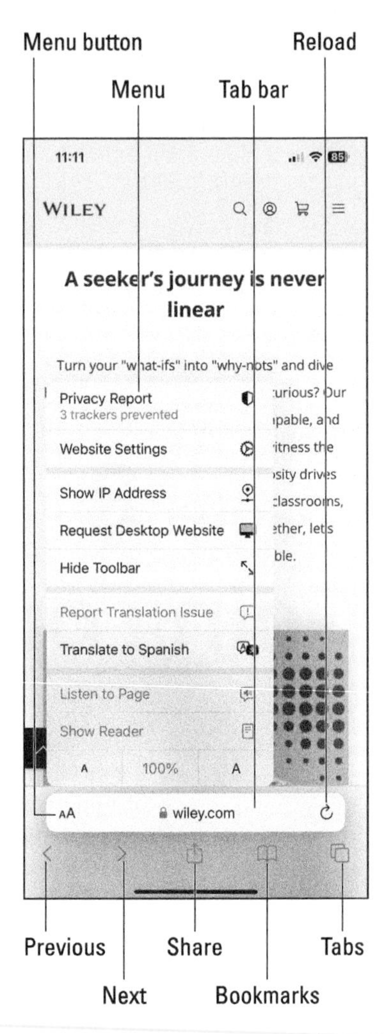

FIGURE 8-1: The iPhone's Safari browser.

TIP

If you prefer a single address bar at the top of the page, choose Settings ➪ Safari, go to the Tabs box, and then select the Single Tab option instead of the Tab Bar option.

You can tap the menu icon (AA) to open the menu, which includes commands for enlarging or reducing the text size, zooming in or out, displaying Apple's Privacy Report, configuring settings for the current website, and requesting the desktop version of the current website.

At the bottom of the screen are the previous and next icons for navigating among the pages you've visited in this tab, the share icon, the bookmarks icon, and the tabs icon for organizing tabs (more on this soon).

REMEMBER

The arrows on the previous icon and next icon appear in blue when there are previous pages or next pages to navigate to. When there are no pages, the arrows appear in gray.

As soon as you start typing in the smart search field, Safari displays a list of web addresses that match those letters. For example, if you type *st*, you might see web listings for the Apple Store (`store.apple.com`), as in Figure 8-2. You may need to scroll to see more suggestions; when you do, the on-screen keyboard hides to make more room for the suggestions.

Safari on your iPhone may suggest web pages you've bookmarked in Safari on your iPad or Mac and synced via iCloud. See the section "Book(mark) 'em, Danno," later in this chapter, for more on bookmarks.

Safari may also suggest web pages from your history list, which logs the pages you visit. See the section "Letting history repeat itself," also later in this chapter.

After all that preamble, you're likely ready to open a web page. Do the following:

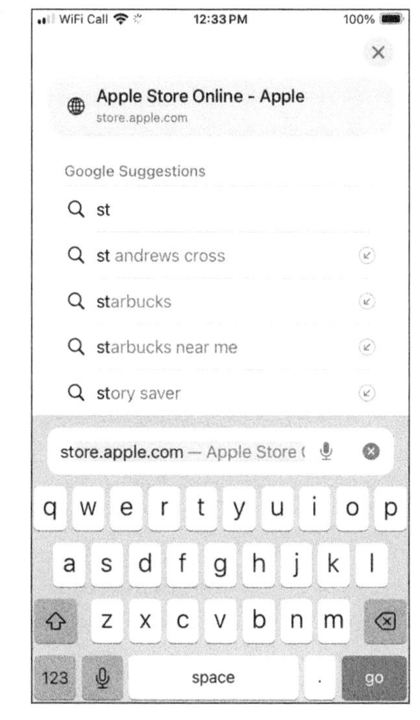

FIGURE 8-2: Web pages that match what you typed.

1. **Tap the Safari icon on the Home screen.**

   It's on the dock by default.

2. **Tap the smart search field in the tab bar to display the keyboard.**

3. **Begin typing the web address on the keyboard.**

   The web address is also called the *URL* (Uniform Resource Locator).

4. **Do one of the following:**

   - *To go to one of the sites on the list, tap the name.* Safari automatically fills in the URL in the address field and displays the page.
   - *Type the rest of the web address, and then tap Go on the keyboard.*

     You don't need to type *www* at the beginning of a URL. If you want to visit `www.theonion.com`, for example, typing `theonion.com` will get you to the satirical website.

TIP

To erase a URL you're typing, tap the circled X to the right of the smart search field, or tap the microphone icon to enter a URL by voice.

The smart search field and other menu options disappear when you start scrolling to read a page. In portrait mode, the URL at the bottom is still visible but it shrinks. In landscape mode, after the smart search field disappears, you don't even see the URL because the site you're visiting claims the entire page. Either way, you get to see more of the page's content.

DESKTOP VERSUS MOBILE WEBSITES

If seeing the Request Desktop Website option in Figure 8-1 raised your eyebrows — some websites have separate versions for "desktop" computers and "mobile" computers. Those terms are in quotes because a desktop computer means either a desktop computer in the conventional sense (usually with a separate screen, keyboard, and mouse or trackpad) or a laptop. A mobile computer means a smartphone or a tablet, with a smaller screen, not a laptop.

If Safari on your iPhone shows you the mobile version of a website, you can request the desktop version by tapping the menu icon, and then tapping Request Desktop Website. (And you can switch back by tapping the menu icon and then tapping Request Mobile Website.)

These days, many websites use a design approach called *responsive design* that enables web pages to adapt intelligently to suit the device on which they're being displayed. Responsive design is gradually reducing the number of websites that have separate desktop and mobile versions.

Browsing the Smart Way

Safari enables you to zoom in on pages so that you can see and read them better. Try these moves:

- **Double-tap the screen so that the portion of the text you want to read fills up the entire screen.** It takes just a second before the screen comes into focus. See Figure 8-3 for an example. It shows two views of the same Wikipedia web page, zoomed out on the left and zoomed in on the right.
- **Pinch the page.** Sliding your thumb and index finger together and then spreading them apart (or, as I like to say, *unpinching*) also zooms out and in of a page.
- **Place two fingers on the page, and then drag it in all directions to pan; flick through a page from bottom to top to scroll down.**
- **Rotate the iPhone to its side.** Landscape orientation gives you a widescreen view with larger text but fewer lines.
- **Pull down from the top of a web page to refresh the page.** Alternatively, tap the reload icon in the tab bar.

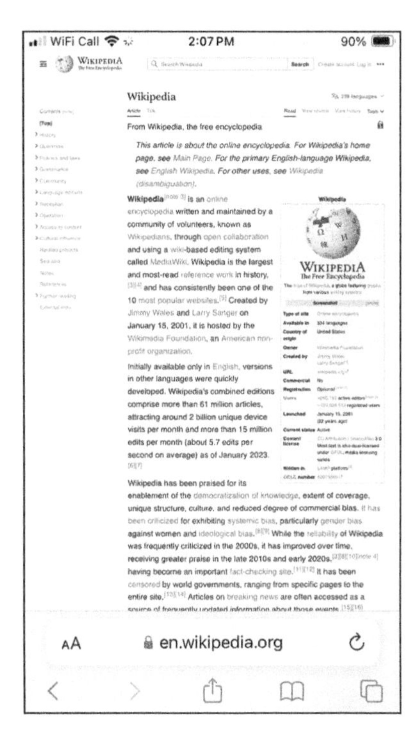

FIGURE 8-3: You can zoom in on a web page to make it easier to read.

Opening multiple web pages at a time

Safari on the iPhone lets you open multiple pages simultaneously. Switching from one open web page to another is now as simple as swiping left or right on the tab bar at the bottom of the screen. (If you move the tab bar to the top of the screen, you can't switch this way.)

Starting up the start page

If the web page you're presently viewing is the only one open, swiping from right to left on the tab bar displays the customizable Safari start page. Favorites and Frequently Visited sites get top billing on this page, followed by other sites shared with you through Messages or Mail, as shown in Figure 8-4.

Scroll down farther on the start page, and you'll see the Privacy Report, which tells you how many trackers Safari has prevented from profiling you. Privacy Report is followed by Siri Suggestions, a reading list of web pages you want to check out later, plus a list of web pages open on your other Apple devices signed into your iCloud account, through what Apple refers to as *iCloud tabs.*

That's the default start page. You can customize it by tapping the Edit button at the bottom of the page, and then working on the Customize Start Page screen (see Figure 8-5). Set the Use Start Page on All Devices switch on (green) if you want the same start page on all your devices. Next, in the main box, set the switch for each item — Favorites, Frequently Visited, Shared with You, Privacy Report, Siri Suggestions, Reading List, Recently Closed Tabs, and iCloud Tabs — on (green) or off (white) to specify which items to show and which to hide. Use the handles on the right side to drag the items into your preferred order. Lastly, set the Background Image switch on (green) to use a background image for the start page; you can then tap an image or tap + (add) to use one of your own images from the Photos app.

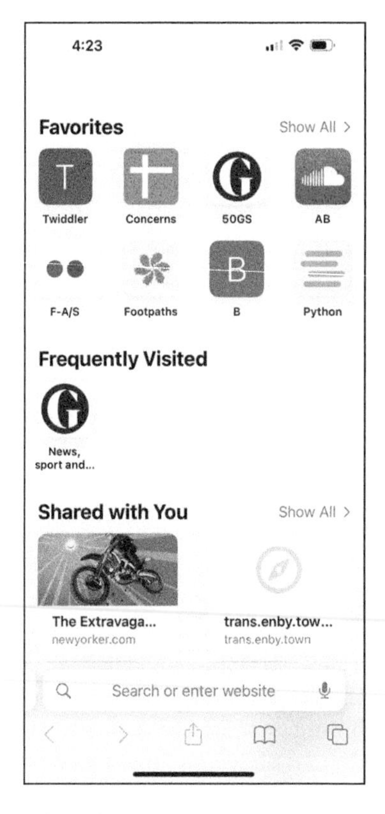

FIGURE 8-4: Scrolling down the Safari start page.

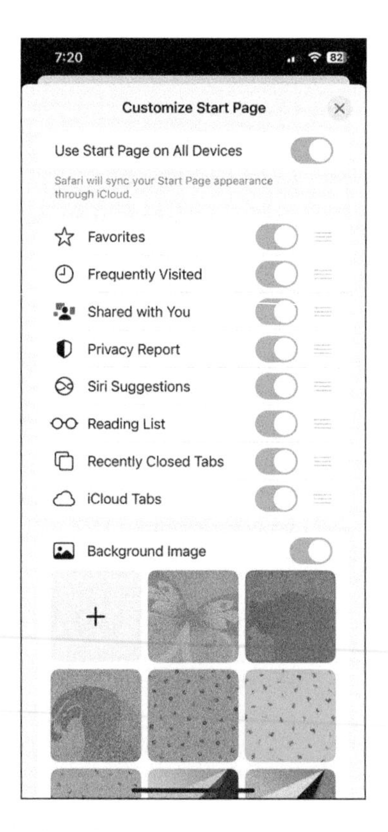

FIGURE 8-5: You can customize the Safari start page.

Keeping tabs on tabs

To see which tabs and web pages you have open in Safari, tap the tabs icon in the lower-right corner (labeled in Figure 8-1). The tabs overview page appears (see Figure 8-6), showing you thumbnails of all the web pages you have open in the active tab group. Each tab group appears as a button on a scrollable button bar across the bottom of the screen, with the active tab group positioned centrally and its name shown in black font rather than gray font. You can switch to another tab group by scrolling the bar sideways and the tapping the tab group's button. Once you've displayed the tab group you want, you can tap a thumbnail to display that web page.

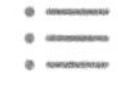

To work with your tab groups, tap the tab groups icon (shown in the margin), and then use the buttons in the Tab Groups pane (see Figure 8-7). You can then:

- **Create a new empty tab group:** Tap the New Empty Tab group button to display the New Tab Group dialog, type the name for the new tab group, and then tap the Save button.
- **Create a new tab group containing tabs:** If the unnamed tab group isn't selected, select it: Tap the top button (which shows 3 Tabs in the figure) to make that tab group active. The Tab Groups pane closes, so tap the tab groups icon again to reopen the Tab Groups pane, this time with that tab group selected. Now that the unnamed tab group is selected, tap the New Tab Group with *n* Tabs button (where *n* is the number). Type the name for the group in the New Tab Group dialog, and then tap the Save button.
- **Rename a tab group:** Swipe left on its button, and then tap the edit icon (pencil) that appears. Type the new name in the Rename Tab Group dialog, and then tap the Save button.
- **Delete a tab group:** Swipe left on its button, and then tap the delete icon (trashcan) that appears.

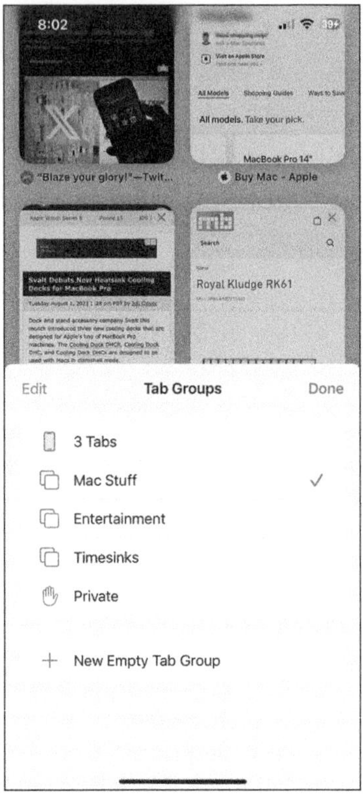

FIGURE 8-6:
The tabs overview page.

FIGURE 8-7:
Lump like-minded web pages into tab groups.

- **Change the order of tab groups:** Tap the edit icon, and then drag a tab group up or down by the handle (three horizontal lines) on its right side. If there's an unnamed tab group, that remains first in the list; you can't move it. Nor can you move the Private tab group, which remains last in the list.
- **Share a tab group:** Tap the edit icon, tap the more icon (ellipsis-in-a-circle) to display the More menu, and then tap Share. In the Share sheet, select the means of sharing (such as Messages) and the contact.

When you finish working in the Tab Groups pane, tap Done to close it.

To close an open web pages, tap the *X* in the upper-right corner of its thumbnail. Alternatively, swipe the thumbnail off the screen to the left.

To add a new web page, tap + in the lower-left corner of the screen, and then select a web destination like you always do.

Looking at links

Surfing the web would be a real drag if you had to enter a URL every time you want to navigate from one page to another. That's why links and bookmarks are so useful. Because Safari functions on the iPhone in the same way browsers work on your PC or Mac, links on the iPhone behave the same way.

Text links are underlined or appear in a different color from other text on the page. Tap the link to go directly to the link's destination.

Long-press a link to display a contextual menu. If the link goes to a web page, the menu typically includes the following commands: Open, Open in New Tab, Open in Tab Group, Download Linked File, Add to Reading List, Copy Link, and Share. To help you decide what to do, Safari displays a preview of the linked site.

Other types of links let you take other actions — for example:

- **Open a map.** Launches the Maps app (see Chapter 12) or Google Maps.
- **Prepare an email.** Tap an email address link to start a new message to that address in the Mail app (discussed in Chapter 10). Long-press an email address link to display the contextual menu, which contains commands such as Send Message (in the Messages app), FaceTime, FaceTime Audio, Add to Contacts, and Copy Email.
- **Make a phone call.** Tap a phone number embedded in a web page, and the iPhone offers to dial it for you. Tap the Call *number* button to place the call. Long-press a phone number to display the contextual menu, whose commands include Send Message (again, in Messages), FaceTime, FaceTime Audio, Add to Contacts, and Copy Phone Number.

Book(mark) 'em, Danno

When you find a web page you'd like to revisit, bookmark it like this:

1. **With the page displayed in Safari, tap the share icon (shown in the margin).**

   The Share sheet opens (see the left screen in Figure 8-8), presenting a slew of sharing-related commands, some built into iOS and some from other apps installed on your iPhone.

2. **Tap Add Bookmark to display the screen shown on the right in Figure 8-8.**

   Safari enters a default bookmark name and suggests your current bookmarks folder (the Favorites folder in the example).

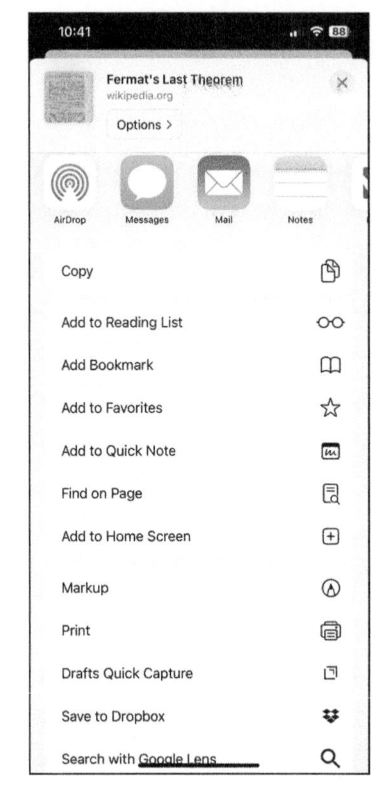

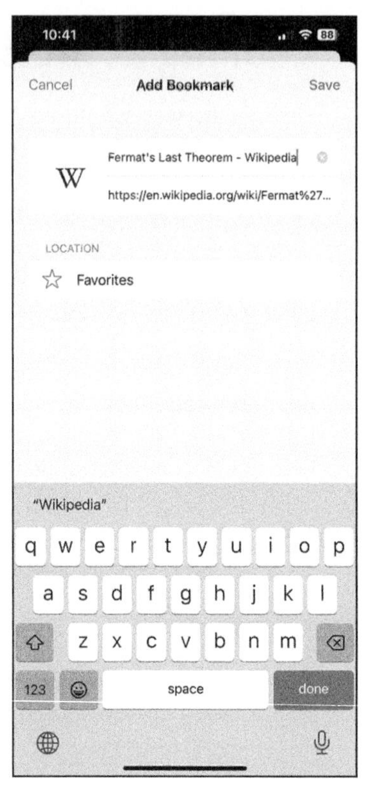

FIGURE 8-8: Creating a bookmark for a web page you want to be able to revisit easily.

3. **Change the bookmark name, as needed.**

   Edit the existing name, or tap the *x*-in-a-circle to delete the name and then type a new one.

4. **Change the location, as needed, by tapping the Location field, and then tapping the folder you want to save the bookmark in.**

5. **Tap the Save button, in the upper-right corner.**

   Safari saves the bookmark and closes the Add Bookmark screen.

Here's how to return to a page you've bookmarked:

1. **Tap the bookmarks icon at the bottom of the screen to display the pane shared by Bookmarks, Reading List, and History.**

2. **If the pane's title bar says Reading List or History, tap the Bookmarks icon on the left to display the Bookmarks tab.**

3. **Tap the folder that contains the bookmark, and then tap the bookmark.**

   Safari displays the bookmarked page.

From the Share sheet, you can tap the following buttons instead of tapping Add Bookmark:

- **AirDrop:** Share the page with users of Macs, iPhones, and iPads with AirDrop turned on.
- **Messages:** Send a link to the web page in a text or an iMessage.
- **Mail:** Start a message in the Mail app containing a link for the page. Mail puts the page's title in the Subject line.
- **Reminders:** Create a reminder in the Reminders app based on this page.
- **Add to Quick Note:** Opens the New Quick Note screen and pastes an image of the web page on it. You can tap the image to go the web page.
- **X:** Add the web page's address to an outgoing tweet on X, The App Formerly Known as Twitter (TAFKAT).
- **Facebook:** Post the page — and whatever comments you choose to add — to the popular social network.
- **Books:** Add the bookmarked page to the Books app.
- **Copy:** Copy the page's URL to the clipboard for pasting elsewhere.
- **Translate:** View the page in another language.
- **Add to Favorites:** Add the page to your favorites for quick access.
- **Add to Reading List:** Add the page to your Reading List, which you can access by opening the Bookmarks pane and tapping the reading list icon (shown in the margin).
- **Add to Home Screen:** Add the site's icon to your Home screen so you can quickly access the site. You can label the Home screen icon before tapping Add to complete the process.
- **News:** Add an article to the News app. This command works only on sites that Apple has partnered with on News. For other sites, the News app opens and displays the Story Unavailable dialog.
- **Print:** Open the Options dialog for printing to an AirPrint printer.
- **Find on Page:** Display the search field and the virtual keyboard. Type your search term, and matches appear highlighted. Tap the up and down arrows to move the yellow highlight through the instances of the term. Tap Done when you're done.
- **Markup:** Create a PDF version of the page and display the markup tools so you can annotate it.

TIP

You'll also see an Options button when you tap the share icon. Tap Options to choose whether to send the web page as a PDF, Reader PDF, or Web Archive.

Altering bookmarks

If you no longer need a bookmark, you can change or get rid of it:

- **To remove a bookmark or folder:** Swipe the bookmark from right to left so that a red Delete button appears on the right, and then tap Delete. Alternatively, swipe all the way to the left so that you don't need to do the extra step of tapping Delete. You can also tap Edit at the bottom-right corner of the screen, tap the red circle next to a bookmark or folder, and then tap the Delete button.
- **To change a bookmark's name or location:** Tap Edit, and then tap the bookmark. The Edit Bookmark screen appears with the name, URL, and location of the bookmark. Tap the fields you want to change. In the Name field, tap the *X* in the gray circle, and then type a new name. Tap the Location field, and then tap the folder to which to move the bookmark.
- **To create a folder for your bookmarks:** Tap Edit, and then tap New Folder in the bottom left. Enter the name of the new folder and choose where to store it.
- **To move a bookmark up or down a list:** Tap Edit, and then drag the three bars to the right of the bookmark's name.

Viewing pages open on your other devices

Your iPhone and other Apple devices sync your open pages and iCloud tabs via your iCloud account. To view another device's pages, open a tab to your start page, go to the From *Device* list or lists, and then tap the button for the page you want to view.

Letting history repeat itself

When you want to revisit a web page you didn't bookmark, find the page in your history.

Safari records the pages you visit and keeps the logs on hand for several days. Tap the bookmarks icon (shown in the margin), and then tap the History tab (the clock icon). Next, tap the day you think you viewed the page. You'll see web pages organized into such categories as This Evening, This Afternoon, and Tuesday Morning. When you find the listing, tap it to display the page.

TIP

To clear your history so that nobody else can trace your steps, tap Clear at the bottom right of the History screen. In the Clear History pane, tap the appropriate button in the Clear Timeframe pane: Last Hour, Today, Today and Yesterday, or All History. Set the Close All Tabs switch on (green) if you want to close all your tabs. Then tap the Clear History button.

You can also clear your Safari history starting from the Home screen. Tap Settings ➪ Safari ➪ Clear History and Website Data and then tap the Clear History button.

WARNING

Clearing your history removes the history information from your iPhone, but your network's administrator (for example, at work) and your ISP will still have a full record of your browsing.

Launching a mobile search mission

To find information, you'll likely want to search the web often. Safari's smart search field makes searching easy and effective. The smart search field is hooked up to a search engine, so you don't need to go to `google.com`, `bing.com`, or another search engine to perform a search.

Configure search settings in Safari

Before searching extensively in Safari, spend a minute choosing search settings. From the Home screen, choose Settings ➪ Safari to display the Safari screen in the Settings app. You can then configure the following settings:

- **Search Engine:** Choose your main search engine, such as Google or DuckDuckGo.
- **Private Search Engine:** Choose the search engine to use for searches in private mode. Tap the Use Default Search Engine button to use your main search engine.
- **Search Engine Suggestions:** Set this switch on (green) to have Safari include suggestions from the search engine. These are usually helpful.
- **Safari Suggestions:** Set this switch on (green) to include suggestions from Safari.
- **Quick Website Search:** Tap this button to display the Quick Website Search screen, and then set the Quick Website Search switch on (green) to enable yourself to restrict a search to a particular website by typing its name in the search. For example, you could type *iMac* *`apple.com`* in the search box to search for iMac information only on `apple.com`.

- **Preload Top Hit:** Set this switch on (green) to have Safari preemptively download the page linked to the top search result so the page appears more quickly if you tap that search result.

Searching in Safari

When you need to search the web, tap the smart search field, and then start typing your search term. Results appear as you type, broken up into categories such as Google Suggestions; Bookmarks, History, and Tabs; and On This Page. Tap the search result whose page you want to view.

Saving pictures from the web

To copy an image from a website, long-press the image so that the pop-up menu appears. You can then tap Save to Photos to save a copy of the image to the All Photos album in your Photos app; tap Copy to copy the picture so you can paste it elsewhere; or tap Copy Subject to have iOS copy the image's subject without the background, again so that you can paste it elsewhere. Copy Subject often works impressively well, but occasionally a human subject's limb goes missing.

Reading clutter-free web pages

Safari's reader view helps you focus on reading a web page by eliminating distractions such as ads. To use reader view, tap the AA icon on the tab bar, and then tap Show Reader from the menu. If the Show Reader command is dimmed on the menu, the web page is not compatible with reader view. Figure 8-9 shows the same Wikipedia article in regular view (left) and reader view (right). To return to the regular view, either long-press the AA icon or tap the AA icon, and then tap Hide Reader.

TIP

To try using reader view all the time, visit Settings ➪ Safari ➪ Reader, and then set the All Websites switch on (green) on the Reader screen.

Translating web pages

To translate a web page, tap the AA icon in the tab bar, tap Translate Website on the menu, and then tap the language you want in the dialog that opens. The translation appears quickly. To go back, tap the AA icon again, and then tap View Original on the menu.

If the Enable Translation? dialog opens, tap the Enable Translation button.

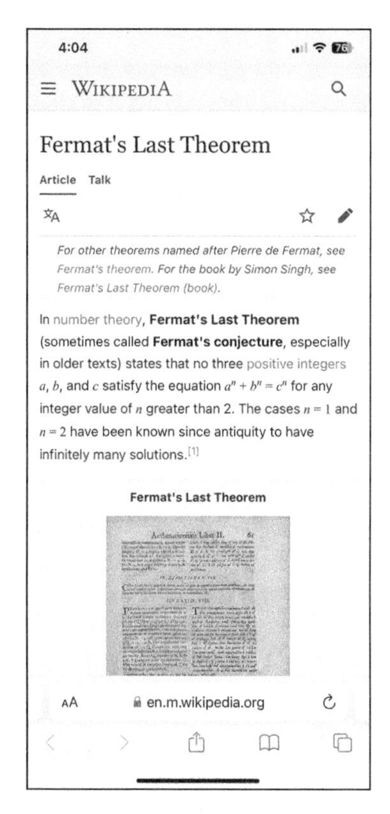

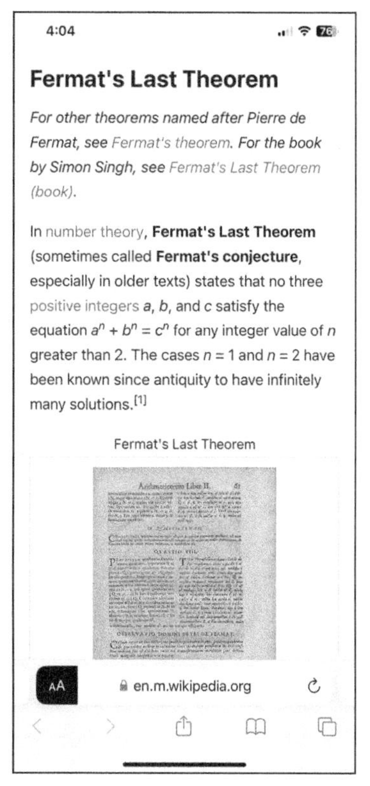

FIGURE 8-9: Use reader view (right) to reduce clutter when reading a web story.

REMEMBER

You can highlight any text anywhere on your iPhone, Safari included, and tap Translate to read in another available language. Besides English, languages include Spanish, Simplified Chinese, French, German, Russian, and Brazilian Portuguese.

Private Browsing

Don't want to leave any tracks while you surf? Turn on private browsing. Safari implements this privacy feature as a tab group called Private that works in a similar way to the tab groups you met earlier in this chapter. The difference is that, while you are browsing in the Private tab group, Safari retains details of the pages you visit and the searches you perform only temporarily, getting rid of them at the end of the private browsing session instead of adding them to your history. (Bear in mind that your network administrator and ISP may still be able to track all your actions.) To further protect you, Safari also prevents advertisers from collecting your iPhone's unique characteristics, which means such companies can't identify your specific device to use in targeted advertising.

To use Private browsing, tap the tab icon so that you see a tabbed view of open pages, and then tap the tab groups icon (shown in the margin). In the Tab Groups pane, tap the Private button. You can then browse as usual. The tab bar and smart search are shaded dark as a visual reminder that you are using private mode.

For security, private browsing locks automatically when your iPhone locks. After you unlock your iPhone, tap Unlock on the Private Browsing Is Locked screen, and then authenticate yourself via Face ID or Touch ID, as appropriate.

Using Sign In with Apple

To help you avoid sharing your email address with companies or people you don't trust, Apple provides the Sign In with Apple feature. Sign In with Apple leverages your Apple ID to enable you to sign in to an app or a website via a unique, random ID instead of your email address. When an app or a website requires an email address, Apple creates a unique custom email address that forwards messages to your actual email address. The random ID and custom email address are linked to your Apple ID.

When you see the Sign In with Apple button in an app (see the left screen in Figure 8-10) or on a web page, click it. On the Sign In with Apple screen (Figure 8-10, middle), click the Continue button. On the next screen (Figure 8-10, right), select the Share My Email option button or the Hide My Email button, as needed. Then click the Continue button and complete the sign-up process in the app or on the web page.

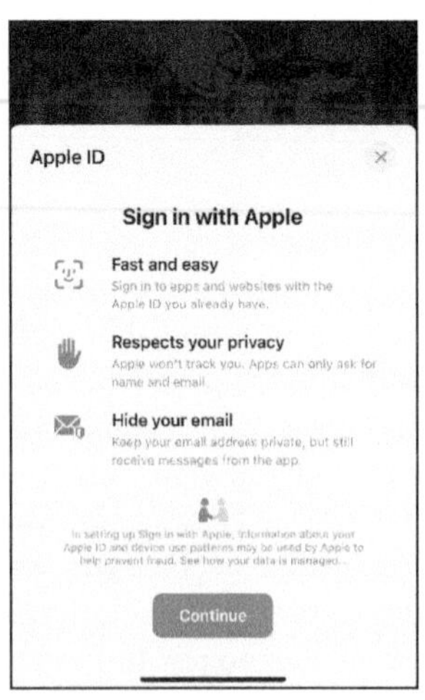

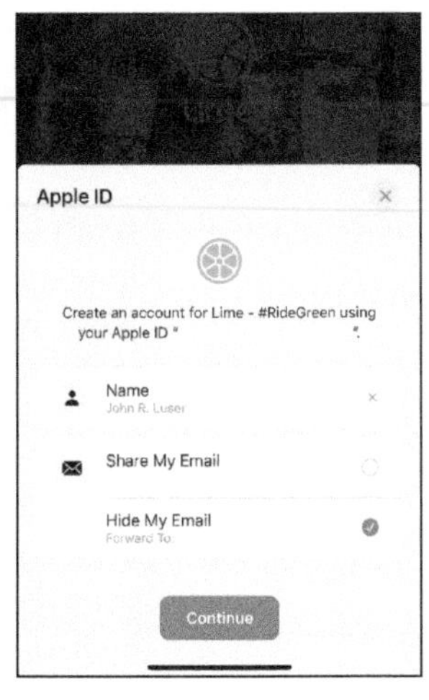

FIGURE 8-10: Starting the Sign In with Apple process in the Lime app.

Creating Multiple Profiles for Safari

NEW

In iOS 17, Safari introduces a feature called Profiles to enable you to cordon off different types of browsing into different silos and to help boost your concentration and privacy. For example, you might create a Work profile to separate your work browsing from your home browsing. Or you might create a profile for each major project you work on, if keeping those projects separate from each other is helpful.

Profiles work as an extension of tab groups. Each profile maintains a separate set of history, cookies, and website data, so (for example) your work history does not get contaminated by your home history, and so on.

Creating a profile

Safari starts you off with a default profile called Personal but keeps it hidden until you create another profile. At that point, the Personal profile pops out of the woodwork, enabling you to switch between it and the new profile you created.

To create a profile, follow these steps:

1. **Choose Settings ⇨ Safari to display the Safari Settings screen.**
2. **Scroll down to the Profiles box, and then tap New Profile to display the New Profile screen (see Figure 8-11).**
3. **Type a descriptive name for the profile, such as *Work*.**
4. **Tap a symbol in the symbol line to represent the profile.**

   You can tap the ellipsis (. . .) at the end of the symbols to display a panel containing other symbols.
5. **Tap the color to use for the profile.**

   You can tap the ellipsis (. . .) to display a panel containing other colors. Safari uses this color on the profile's start page to help you identify the tab group.

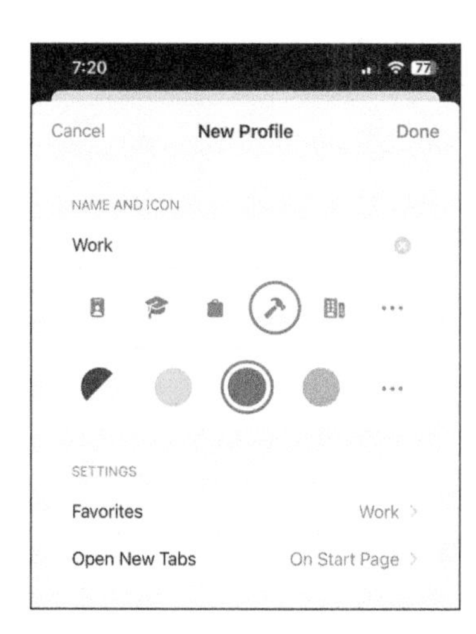

FIGURE 8-11: Enter the profile's name and details on the New Profile screen.

6. **Tap the Favorites button, and then tap the folder in which to store the favorites for this profile.**
7. **Tap the Open New Tabs button, and then select the location for opening new tabs, such as On Start Page.**

8. **Tap the Done button in the upper right to create the profile and close the dialog.**

   The profile appears in the Profiles list on the Safari screen in Settings. The Profiles list now also shows the Personal profile, which had been hidden up till now.

   If you want, you can now create another profile by repeating Steps 2 to 8.

Switching from one profile to another

To switch from one profile to another, you open a tab group for the profile you want to use. Tap the tabs icon to display the overview screen, and then tap the tab groups icon. This icon now bears the graphic assigned to the profile you're currently using instead of the standard tab groups graphic. The Tab Groups pane opens, showing the profile's name — for example, Personal Tab Groups for the Personal profile.

Tap the Profile pop-up menu, and then tap the profile you want to use (see Figure 8-12, left). The start page appears, showing the profile's color and icon (see Figure 8-12, right). You can then open tabs and browse as usual.

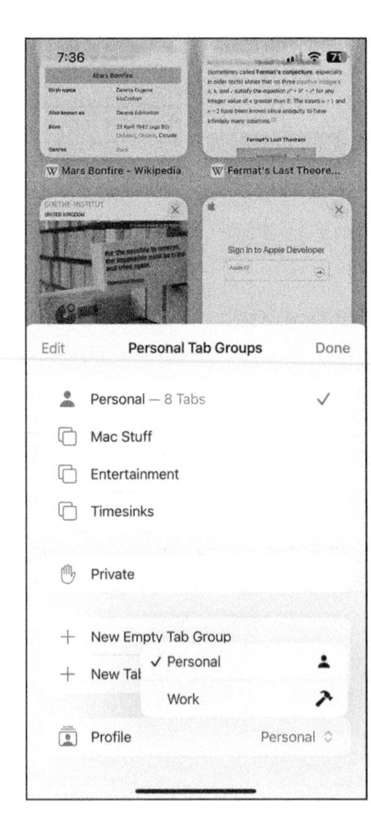

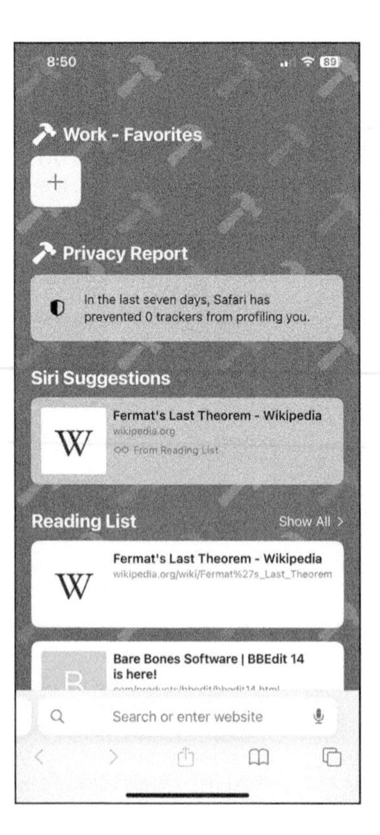

FIGURE 8-12: Use the Profile pop-up menu (left) to open a new tab group for a different profile. The profile's start page uses the profile's color and icon (right).

Deleting a profile

To delete a profile, choose Settings ➪ Safari, tap the profile in the Profiles box, and then tap the Delete Profile button. In the confirmation dialog, tap Delete.

Choosing Smart Safari Settings

To keep your Safari browsing as safe as possible from the myriad threats online, choose Settings ➪ Safari, and then configure the following settings:

- **AutoFill:** When AutoFill is turned on, Safari can automatically fill out web forms by using your personal contact information, usernames, passwords, and credit card information.
- **Favorites:** Designate the folder to treat as a Favorites folder, letting you quickly access favorite bookmarks when you enter an address, search, or open a new tab.

NEW

- **Extensions:** An *extension* is software that adds capabilities to Safari, such as blocking ads or scam callers. Tap More Extensions to install extensions, and then use the controls in the Extensions pane to enable the features you want to use.
- **Open Links:** Decide whether links open in a new tab or in the background.
- **Tabs box settings:** You can choose between positioning the tab bar at the bottom of the screen (the new default) or using with a single tab view at the top. You can also choose whether to use a landscape tab bar and whether to allow website tinting, a design aesthetic. The Close Tabs screen lets you choose whether to close tabs manually or have Safari automatically close tabs you haven't viewed for a day, a week, or a month.
- **Prevent Cross-Site Tracking:** Set this switch on (green) to prevent advertisers from tracking you as you browse across sites.
- **Hide IP Address:** Tap this button, and then tap From Trackers to have Safari block trackers, tools that track your browsing activity.
- **Require Face ID/Touch ID to Unlock Private Browsing:** Set this switch on (green) to have Safari require you to unlock the Private tab group after your iPhone locks.
- **Fraudulent Website Warning:** Set this switch on (green) to have Safari warn you when you happen upon a website known to try to steal visitors' user-names, passwords, and so on or to try to install malicious software on their devices.

- **Settings for Websites:** In this box, set the Share Across Websites switch on (green) to have Safari share these settings across all your iCloud devices. Then tap the Page Zoom button, the Request Desktop Website button, the Reader button, the Profiles button, the Camera button, the Microphone button, and the Location button, in turn, and choose the default settings you want to use for these features on all websites.
- **Automatically Save Offline:** Set this switch on (green) to have Safari automatically save all your Reading List items locally so that you can read them offline.

Tap the Advanced button to display the Advanced screen, where you can configure the following settings:

- **Website Data:** Tap this button to display the Website Data screen, where you can see which websites have stored data on your iPhone. You can delete a website's data by swiping left on it, and then tapping Delete.
- **Advanced Tracking and Fingerprinting Protection:** Tap this button, and then choose Off, Private Browsing (the default), or All Browsing to tell Safari when to use its Advanced Tracking and Fingerprinting Protection feature.
- **Block All Cookies:** Set this switch on (green) if you want Safari to block all cookies, the tiny tracking files that websites place to recognize you and store your browsing activity.

WARNING

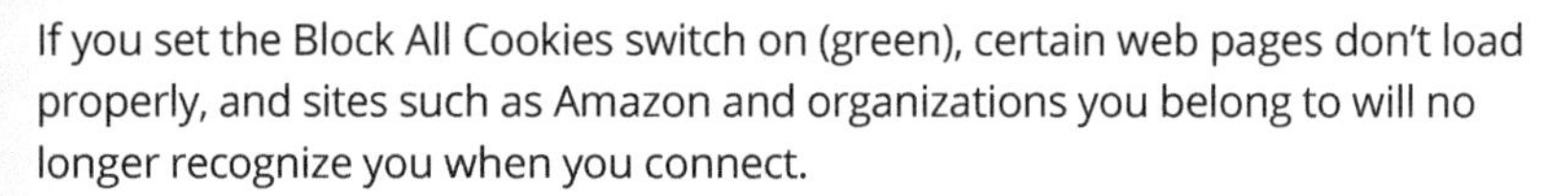

If you set the Block All Cookies switch on (green), certain web pages don't load properly, and sites such as Amazon and organizations you belong to will no longer recognize you when you connect.

- **Privacy Preserving Ad Measurement:** Set this switch on (green) to allow Apple to measure the performance of advertisements while preserving your privacy.
- **Check for Apple Pay:** Set this switch on (green) to allow websites to determine whether you use Apple Pay.
- **JavaScript:** Set this switch on (green) to allow websites to use JavaScript, a programming language that's widely used to add functionality to web pages. JavaScript is completely separate from Java, a more powerful programming language notorious for security problems.
- **Web Inspector:** Set this switch off (white) unless you're a developer.
- **Remote Automation:** Set this switch off (white) unless you're a developer.
- **Feature Flag:** This button leads to a screen of developer-only features. Leave them alone.

IN THIS CHAPTER

- » Sending and receiving SMS and MMS messages
- » Sending and receiving voice and video iMessages
- » Using the Notes app

Chapter 9

Texting 1, 2, 3: Messages and Notes

In this chapter, you come to grips with Messages, Apple's instant-messaging app, and Notes, its note-taking and sharing app. Messages lets you communicate using the iPhone, the iPad, the Mac, and Apple Watch. Notes enables you to take notes on the iPhone, the iPad, and the Mac.

Getting the iMessage

The Messages app lets you exchange short text messages with any cellphone that supports the SMS (Short Message Service) protocol. You can also send and receive MMS (Multimedia Messaging Service) messages, which can include pictures, contacts, videos, audio recordings, and locations, with any cellphone that supports the MMS protocol.

Beyond SMS and MMS, which your cellular provider offers, your iPhone also supports the iMessage protocol. iMessage uses an Apple-provided service to send and receive unlimited messages with text, pictures, contacts, videos, and locations without going through your carrier.

The good news is that although the wireless operators all charge *something* for SMS or MMS services, iMessages are free as long as you send and receive them over Wi-Fi. If you use cellular service, there's no per-message charge, but an iMessage does use cellular data from your plan's allotment.

The bad news is that you can share iMessages only with people using iPhones, iPads, Macs, and Apple Watch. And, if no data connection is available when your iPhone is using a cellular connection rather than Wi-Fi, the message will be sent as a standard SMS or MMS message via your wireless carrier (and subject to the usual text or multimedia message charges where applicable).

If you have a limited data plan, choose Settings ⇨ Cellular to display the Cellular screen, and then set the Cellular Data switch off (white) to force Messages (and all other data-using apps) to use only Wi-Fi. Alternatively, go to the Cellular Data box toward the bottom of the Cellular screen, and then set the switch off (white) for each app you don't want to use cellular data. You can also turn off the Send as SMS feature if you like, which may be a good idea if you don't have unlimited messages as part of your data/voice package. To do that, choose Settings ⇨ Messages, and then set the Send as SMS switch off (white).

Before you send or receive messages, understand these messaging essentials:

- **Most mobile phones can send and receive SMS messages; many can send and receive MMS messages as well.** The iPhone is the only mobile phone that can send iMessage messages.
- **Some phones limit SMS messages to 160 characters.** If you try to send a longer message to one of these phones, your message may be truncated or split into multiple shorter messages, with each counting as one SMS message unless you have an unlimited plan. Good news: Your iPhone is smarter than this.

  The Messages app can count characters for you. To enable this feature, choose Settings ⇨ Messages, and then set the Character Count switch on (green). Now when you type more than a line of text, you'll see the number of characters you've typed so far, a slash, and then the number 160 (the character limit on some mobile phones, as just described) directly above the Send button.

  Even if you enable the Character Count feature, Messages doesn't display the character count for iMessages, because these have no length limit.
- **No matter which carrier you choose, you'll encounter a bewildering array of pricing options.** If you don't subscribe to a messaging plan from your wireless operator — either a stand-alone plan or in a bundle — you'll pay up to 20¢ or 30¢ per SMS or MMS message sent or received.

WARNING

Each individual message in a conversation counts against this total, even if it's only a one-word reply such as "OK."

- **SMS and MMS messages require access to your wireless operator's cellular network.** iMessages, on the other hand, can be sent and received over either a cellular network or a Wi-Fi network.

Sending text messages

Tap the Messages icon on the Home screen to launch the Messages app, and then tap the new message icon (shown in the margin) in the top-right corner of the screen to start a new message.

TIP

You can long-press the Messages icon on the Home screen and use the quick action to open the app directly to any of the three most recent conversations or to begin a new blank message. And if you're not in the habit of long-pressing Home screen icons, you should be — many offer quick action shortcuts that let you open the app and perform an action with just a single long-press.

At this point, the To field is active and awaiting your input. You can do three things at this point:

- If the recipient isn't in your Contacts list, type the person's cellphone number or email address (iMessage users only).
- If the recipient is in your Contacts list, type the first few letters of the name. A list of matching contacts appears. Tap the contact's name, making Messages enter it as a button. You can then begin typing another name to send the message to multiple recipients at once.
- Tap the blue + icon on the right side of the To field to select a name from your Contacts list.

TIP

You can also compose the message first and address it later, which can help you avoid sending the message before it's complete. Tap inside the text-entry box (the white rounded rectangle above the keyboard) to activate it, and then type your message. When you've finished typing, tap the To field and enter the address.

When you've finished addressing and composing, tap the send icon (shown in the margin) to send your message. The circle is blue when you're sending an iMessage but green when sending an SMS or MMS.

Congratulations — you've sent a text message. Read over your wise words and smile.

Uh-oh. The message contains a mistake — a dubious double-entendre, an insulting innuendo, a phonetic faux pas, or something worse and lacking alliteration. What to do? If the message is an SMS or MMS, there's nothing you can do to get it back. Move on (or start dusting off your excuses — or your resume).

But if the message is an iMessage, you can edit it or undo sending it. Either way, long-press the message to display the menu shown in Figure 9-1. Then tap Undo Send to recall the message; or tap Edit to open it for editing, type your correction, and then tap the correct icon (shown in the margin).

FIGURE 9-1: Messages enables you to edit a message you've sent or even undo the sending.

WARNING

The Undo Send command is available for only 2 minutes after you send the message, so you need to move quickly.

REMEMBER

Editing and un-sending messages works only on iOS 16 and later versions, iPadOS 16 and later versions, and macOS Ventura and later versions. Even with these versions, the recipient may see the message before you undo the send. If you edit the message, the recipient sees the message marked as *Edited To* and the edited version, but they can also display the original version if they want.

Alert: You've got messages

Your iPhone can alert you to new messages with an audio alert, an on-screen alert, or both.

If you want to hear a sound when any message arrives, go to the Home screen and tap Settings ➪ Sounds & Haptics ➪ Text Tone, and then tap one of the available sounds. You can audition any sound in the list by tapping it.

WARNING

You hear the sounds when you audition them in the Settings app, even if you have the ring/silent switch set to silent. After you exit the Settings app, however, you won't hear a sound when a message arrives if the ring/silent switch is set to silent or the Do Not Disturb feature is enabled.

If you don't want to hear an alert when a message arrives, instead of tapping one of the listed sounds, tap the first item in the list: None.

You can also assign a custom alert sound to anyone in your Contacts list. On the Contacts screen in the Phone app or in the Contacts app, tap the contact, and then tap Edit. Tap Text Tone, select the tone, tap Done, and then tap Done again.

TIP

While on the Text Tone screen, you can set the Emergency Bypass switch on (green) to allow this contact's text tone to play when your iPhone is silenced or a focus is on. You can also set Emergency Bypass separately for the contact's ringtone.

In addition to playing a sound when a new message arrives, you can receive on-screen notifications. (Chapter 5 discusses notification settings.)

If you receive too many messages from people or companies you don't know, choose Settings ➪ Messages, and then set the Filter Unknown Senders switch on (green). Next, tap Settings ➪ Notifications ➪ Messages ➪ Customize Notifications, go to the Allow Notifications box, and then set the Unknown Senders switch off (white).

After you set the Unknown Senders switch on, Messages opens to the Filters screen, where you choose what you want to view: All Messages, Known Senders, Unknown Senders, Unread Messages, or Recently Deleted.

Reading and replying to text messages

You can read and reply to a new message notification as follows:

- **If your iPhone is locked when you receive a notification, you can**
  - *Launch the Messages app:* Tap the notification.
  - *Clear the notification:* Slide the notification left, and then tap Clear.
  - *Reply without unlocking your iPhone:* Long-press the notification.
- **If your iPhone is unlocked when you receive a notification, you can:**
  - *Launch the Messages app:* Tap the notification.
  - *Reply in place:* Long-press the notification.

To reply to the message with text, type your reply in the text-entry box and then tap Send.

To read or reply to a message after you've dismissed its notification, tap the Messages icon. If a message other than the one you're interested in appears on the screen when you launch the Messages app, tap the < icon in the top-left corner of the screen, and then tap the sender's name; that person's messages appear on the screen.

TIP

If you don't see a Send button, type one or more characters to make the Send button replace the microphone icon on the right side of the text-entry box.

Long-press a sender's name (on the Messages screen) to peek at a preview of their last few messages. Tap a sender's name to open the message in the Messages app. This technique is great for triaging lots of messages quickly.

TIP

Be sure you're pressing the sender's name and not the circular picture or monogram on the left. Long-pressing the sender's picture or monogram displays options for responding with a message, a phone call, a FaceTime video chat (if available), or an email, or even paying the sender with Apple Pay.

Messages enables you to send a voice message instead of text. Tap + (add icon) to open the Add menu, and then tap the audio icon (shown in the margin) on the Apps bar, tap the red microphone icon, speak your fill (heck, sing if you want), and then tap the red stop button (see Figure 9-2). You can then tap the play button to hear what you just said, swipe upward to send your voice message, or tap the X to delete it.

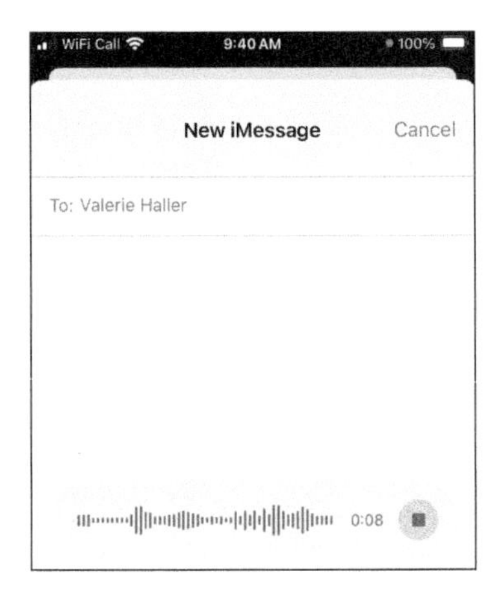

FIGURE 9-2: Tap the audio icon to record a voice message.

You can also tap the microphone icon in the text-entry box or the microphone key next to the spacebar on the keyboard and then dictate your reply to Siri, who will translate it to text and type it in the text-entry box for you. (To find out more about Siri, your intelligent assistant, see Chapter 4.)

In your text conversations, the individual messages appear as a series of bubbles. Your messages appear on the right side of the screen, in blue bubbles if the recipient is using iMessage or green bubbles if the person is using SMS/MMS; the other person's messages appear on the left in gray bubbles, as shown in Figure 9-3.

See how the words *12:29PM tomorrow* have a blue underline, like a link? You can tap underlined words to display additional options, enabling you to create a calendar event, show the date on the calendar, or copy the event to the clipboard (for pasting elsewhere).

You can easily delete individual messages or entire conversations:

FIGURE 9-3: Conversations in Messages look like this. The ellipsis (. . .) indicates the other person is typing.

- **Delete messages from a conversation:** Open the conversation so you can see the messages it contains. Long-press any text bubble, and then tap More. You'll now see an empty check circle to the left of each text bubble. Tap a text bubble that you want to delete, and its circle fills in with blue and displays a white check mark. When you've added a check mark to all the text bubbles you want to delete, tap the blue trashcan icon at the bottom left of the screen.

  TIP

  To forward to another mobile phone or Apple device user all or part of a conversation (as an SMS, MMS, or iMessage message), long-press a text bubble, tap More, and then tap each text bubble that you want to forward. Now tap the blue arrow at the bottom right of the screen. Messages copies the contents of the text bubbles with check marks to a new text message; specify a recipient, and then tap Send.

- **Delete an entire conversation:** Swipe the conversation from right to left, and then tap the red Delete button.
- **Delete multiple conversations:** Tap Edit at the top left of the Messages list, and then tap Select Messages; if you've enabled Filter Unknown Senders, the Edit button doesn't appear, so tap the ellipsis-in-a-circle, and then tap Select Messages. A circle appears to the left of each conversation. Tap each conversation you want to delete, so that Messages selects the conversation's circle, and then tap the Delete button at the bottom right to delete those conversations.

TIP

You can tell Messages to either keep messages forever or automatically delete them after 30 days or one year; to configure this option, tap Settings ➪ Messages ➪ Keep Messages. To review or recover messages that you or Messages have deleted, go to the main Messages screen and tap the Recently Deleted category.

MMS: Like SMS with media

To send a picture or video in a message, follow the instructions for sending a text message, but before you tap Send, tap + (add icon) to the left of the text-entry box, and then tap Camera to shoot a new photo or video, or tap Photos to select an existing photo or video. You can add text to photos or videos (by typing in the text-entry box) before or after you select them. When you're finished, tap the Send button.

If you receive a picture or video (or voice message) in an iMessage, it appears in a bubble just like text. Tap the element to see it full-screen.

TIP

If you enable the Raise to Listen option in Settings ➪ Messages, you can quickly listen to and reply to a voice message by merely raising your iPhone to your ear.

Tap the share icon in the lower-left corner (and shown in the margin) for options including Message, Email, Twitter, Facebook, Save Image, Assign to Contact, Copy, and Print. You'll also see an option to share the photo or video via AirDrop (see Chapter 12).

If you don't see the share icon, tap the picture or video once to summon it.

Tap the round picture icon at the top of a conversation and then scroll down to see every image in this conversation. Tap See All to see all the pictures; swipe left or right, respectively, to see the previous or next images or videos. You can also scroll down a bit farther to see all the links and documents from this conversation.

If the FaceTime camera icon (shown in the margin) appears in the upper-right corner of a conversation, tap it to start a FaceTime audio or video call with that person.

Massive multimedia effects

Multimedia effects — tapbacks, bubble and screen effects, handwriting, annotation, Digital Touch, emoji suggestions, and the built-in App Store for third-party add-ons — enhance your iMessages. Here's the lowdown on these features:

- **To comment on a message with a tapback:** Long-press (or double-tap) the message until the tapback bubble appears, as shown in Figure 9-4.

  If you change your mind and want to dismiss the tapback bubble without adding a tapback icon to the message, just tap the screen anywhere *outside* the tapback bubble.

WARNING

Tapback works only with other devices running iOS 10 or later, iPadOS 13, or macOS Sierra or later. If the recipients are using older versions, they won't see any cute balloons, just plain old text that reads, "*Your-name-here* Loved/Liked/Disliked/ Laughed At/Emphasized/Questioned *item-name-here*."

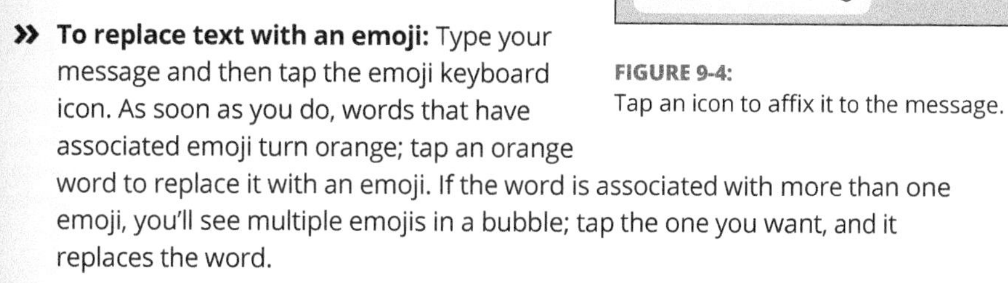

FIGURE 9-4:
Tap an icon to affix it to the message.

- **To replace text with an emoji:** Type your message and then tap the emoji keyboard icon. As soon as you do, words that have associated emoji turn orange; tap an orange word to replace it with an emoji. If the word is associated with more than one emoji, you'll see multiple emojis in a bubble; tap the one you want, and it replaces the word.

- **To add bubble or screen effects:** Prepare your message as usual but rather than tapping the little up-arrow-in-a-circle to send it, long-press it until the Send with Effect screen appears. Tap the Bubble tab at the top to select Slam, Loud, Gentle, or Invisible Ink as the bubble effect for your message. Or tap the Screen tab and swipe left or right to select Echo, Spotlight, Balloons, Confetti, Love, Lasers, Fireworks, or Celebration as the screen effect for this message.

  If you change your mind and don't want to add an effect, tap the *x*-in-a-circle to dismiss the Send with Effect screen.

- **To send a handwritten message:** Sometimes nothing but a handwritten note will do. To send one, first rotate your iPhone 90 degrees so the long edge is parallel with the ground (landscape mode). If the handwriting interface doesn't appear when you rotate the iPhone, tap the handwriting icon in the lower-right corner of the keyboard (and shown in the margin).

  Either tap a preset message at the bottom of the screen to use it or just start writing in the big white area. If you fill the message area with text and want to add more white space, tap > on the right side of the screen; you can then tap < on the left side to go back if you need to.

- **To send Digital Touch effects:** Digital Touch effects were introduced on the Apple Watch and have spread to the Messages app. To use them, tap + (add icon) to display the menu, tap More to expand the menu, and then tap Digital Touch (whose icon is shown in the margin) to reveal the Digital Touch interface shown in Figure 9-5. Then:

  - *To expand the Digital Touch interface to full-screen:* Tap the gray horizontal bar between the App bar and the Digital Touch panel.
  - *To sketch:* Draw with one finger.

- *To send a pulsing circle:* Tap with one finger.
- *To send a fireball (see Figure 9-5):* Press with one finger.
- *To send a kiss:* Tap with two fingers.
- *To send a heartbeat:* Long-press with two fingers.
- *To send a broken heart:* Long-press with two fingers, and then drag down.
- *To switch ink colors:* Tap the big blue circle on the left to reveal additional colors.
- *To add a picture or video to your Digital Touch effect:* Tap the camera icon.

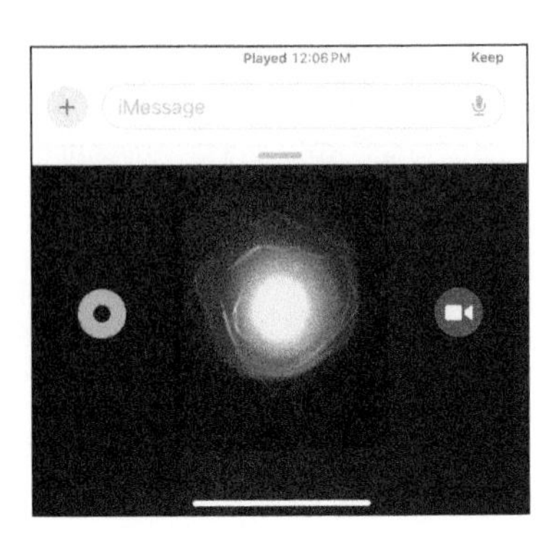

FIGURE 9-5: The Digital Touch interface ready to launch a fireball.

WARNING

Digital Touch messages are sent automatically, so don't tap or press the screen in Digital Touch mode unless you mean it.

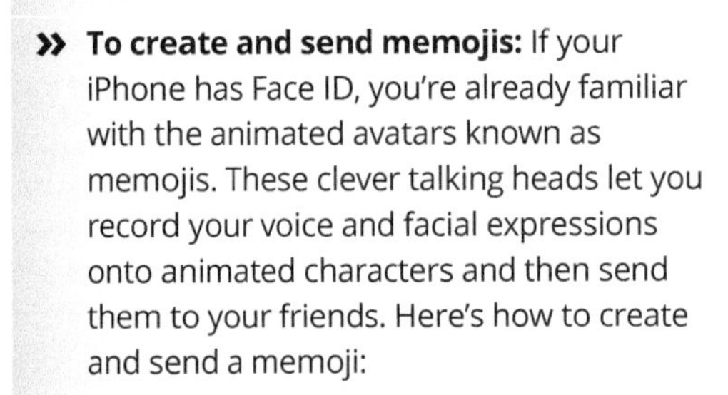

» **To create and send memojis:** If your iPhone has Face ID, you're already familiar with the animated avatars known as memojis. These clever talking heads let you record your voice and facial expressions onto animated characters and then send them to your friends. Here's how to create and send a memoji:

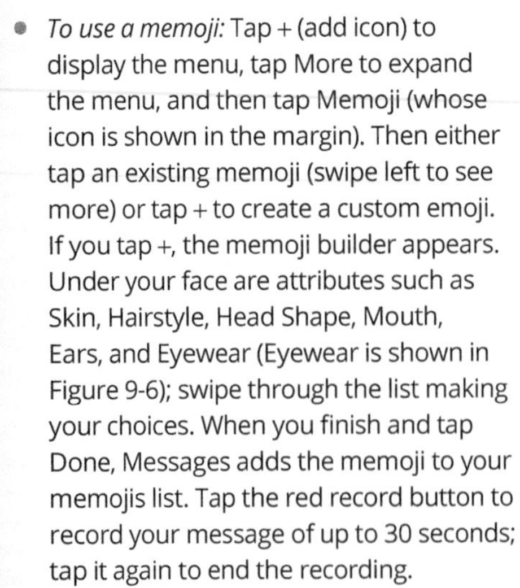

- *To use a memoji:* Tap + (add icon) to display the menu, tap More to expand the menu, and then tap Memoji (whose icon is shown in the margin). Then either tap an existing memoji (swipe left to see more) or tap + to create a custom emoji. If you tap +, the memoji builder appears. Under your face are attributes such as Skin, Hairstyle, Head Shape, Mouth, Ears, and Eyewear (Eyewear is shown in Figure 9-6); swipe through the list making your choices. When you finish and tap Done, Messages adds the memoji to your memojis list. Tap the red record button to record your message of up to 30 seconds; tap it again to end the recording.

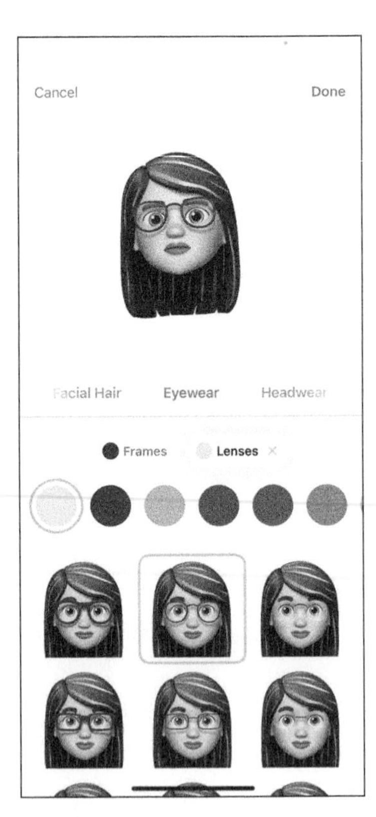

FIGURE 9-6: Choose the features, accessories, and clothing for your memoji.

- *After creating the memoji:* The memoji plays automatically, letting you review your message. You can tap Replay play it again. Tap the send icon to send it, or tap the red trashcan icon to delete it without sending.

Messages has a built-in app store so you can buy (or download for free) sticker packs, new effects, and more. To get some free stickers and see what other third-party add-ons are available, tap + (add icon), and then tap Stickers on the menu. Now swipe the row of sticker apps to the left, tap Edit to display the Manager Sticker Apps screen, and then tap Get Sticker Apps on the App Store.

Once you've acquired sticker apps, tap +, tap Stickers on the menu, and then tap the sticker app in the row of sticker apps to display the available stickers. You can then tap a sticker to add it to your message.

If you use Messages much, pin your favorite conversations to the top of the list for easy access. To pin a conversation, long-press it, and then choose Pin. The conversation moves from the list to a circle at the top of the list. The pinned conversations display icons for tapbacks, as well as new messages and typing indicators.

To unpin a conversation, long-press it and choose Unpin. To unpin multiple conversations, tap Edit at the top left of the Messages list, tap Edit Pins, and then tap the minus-in-a-circle for each conversation you want to unpin. (If you've enabled Filter Unknown Senders, the Edit button doesn't appear; instead, tap the ellipsis-in-a-circle, and then tap Edit Pins.) You can quickly pin an unpinned conversation by tapping the pin in a yellow circle to its right. Tap Done when you're done.

SharePlay via Messages

If you've used Apple's SharePlay feature via FaceTime (see Chapter 7), you'll appreciate that you can also use Messages to establish a SharePlay session. Just locate the content you want to share in the appropriate app, such as a song in the Spotify app; tap the share icon (arrow escaping a box); and then tap the ellipsis. On the overlay that appears (see Figure 9-7, left), tap the SharePlay button. On the SharePlay overlay, choose the contact with whom to share, and then tap Messages (see Figure 9-7, right). The SharePlay link appears in a new message to the contact. You can type any persuasive text needed or simply send the SharePlay link unadorned. The recipient taps the Join button to establish the connection, and then one of you taps the Start button.

FIGURE 9-7: To use SharePlay via messages, tap the SharePlay button on an app's overlay (left), and then tap Messages on the SharePlay overlay (right).

Group messaging has never been better

If one-on-one conversations in Messages are good, conversing with a group of people is even better.

To start a group conversation, just add two or more people in the To field. Other than that, group conversations work the same way as individual messages, so everything in this chapter works for group conversations, too.

Tap the circular pictures of the participants at the top of any group conversation and then tap the info icon (an *i*-in-a-circle). This brings up the details screen, where you can enter a name for this group chat, initiate a phone call or FaceTime chat with any participant, share your location, and more. When you're in a lively conversation with a handful of friends and need to silence its all-too-frequent notifications, set the Hide Alerts switch on (green) to suppress alerts until you're ready to participate fully again.

To remove yourself completely from a group chat, tap Leave This Conversation.

Send texts by emergency SOS via satellite

If you and your iPhone 14 or iPhone 15 are in trouble somewhere without a phone signal, you can use the Emergency SOS via Satellite feature to send text messages to get help. When the Phone app shows the message *No Connection: Try Emergency Text via Satellite,* tap Emergency Text via Satellite in the bottom-right corner of the screen, and then follow the prompts to specify what's wrong and summon assistance.

NEW

You can use the Roadside Assistance via Satellite to text AAA, the largest roadside assistance provider in the US, for assistance if your vehice breaks down somewhere with no cellular or Wi-Fi coverage. Roadside Assistance via Satellite is available only in the US and is free for two years after you activate any iPhone 14 model or iPhone 15 model.

Taking Note of Notes

Notes is a full-featured note-taking app. In addition to creating notes you save (or share), Notes can also capture maps, photos, and URLs, and let you draw with your finger. The Attachments Browser feature makes it easy to locate those photos, maps, and everything else you attach to your notes.

Start by tapping the Notes icon on the Home screen. Or long-press the Notes icon, and then use the quick action to open directly to a new blank note, checklist, photo, or scan. When the Folders screen appears, tap any folder to reveal a list of its contents.

REMEMBER

You can sync notes with your computer (see Chapter 3). And if you've enabled note syncing for more than one account (tap Settings⇨Mail⇨Accounts), you'll see not only your iCloud Notes folder but also a folder for each additional account that's syncing Notes. Tap a folder to see its list of its contents; tap the < in the upper-left corner of the screen to return to the list of folders.

TECHNICAL STUFF

An *account* refers to a Microsoft Exchange, iCloud, Apple ID, Gmail, Yahoo!, AOL, or other account that offers a notes feature.

Now, create a note by tapping the new note icon (pencil on paper) in the lower-right corner of the screen. The virtual keyboard appears. Type your note. To modify the formatting or spice up your note with a checklist, photo, video, or drawing, use the icons above the keyboard (from left to right): formatting, checklist, table, insert photo or video, and finger drawing. Figure 9-8 shows a checklist and a table. To dismiss the icons and go back to typing text, tap the X on the right side of the row of icons.

When you're finished, tap the Done button in the top-right corner to save the note.

You can also share notes with anyone. To invite someone to share a note, tap the share icon (arrow escaping a box) at the top of the screen. At the top of the overlay that opens, choose either Collaborate or Send Copy in the pop-up menu, and then choose the means of sharing, such as Messages. The invitee receives a link that they can click to open the note.

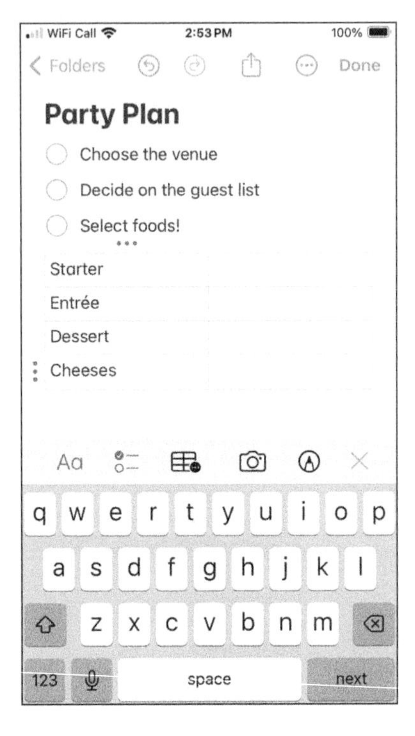

FIGURE 9-8: Creating a note in Notes.

TIP

Tags help you organize and filter notes by category. To add a tag to a note, just type # anywhere in the body of your note, followed by the tag name (for example, #invoices, #useful, #coolstuff) followed by a space or return. To see all notes assigned to a particular tag, tap the tag name in the Tags section near the bottom of the Folders screen. The Tags section appears only when you've added at least one tag to at least one note.

To keep a note at the top of the Notes screen, long-press the note, and then tap Pin Note. The note then appears in the Pinned section.

To delete a note, swipe it from right to left and then tap the trashcan icon.

TIP

The Quick Note feature enables you to start a note from another app or from Control Center. In an app, tap the share icon (arrow escaping a box) to display the overlay, and then tap Add to Quick Note or New Quick Note (some apps have one command; some have the other). For Control Center, first add the Quick Note item by choosing Settings⇨Control Center and tapping Quick Note in the More Controls list. You can then open Control Center and tap Quick Note to display the New Quick Note screen.

Tap the < icon at the top-left corner of the screen to return to the list of notes. From here, you can tap any note to open it for viewing or editing.

TIP

When viewing a notes list, you can tap the ellipsis in the upper-right corner to reach three useful commands: View as Gallery displays your notes as thumbnails; Select Notes enables you to select multiple notes, which you can then move to a different folder or delete all at once; and View Attachments displays Attachments Browser, which rounds up every photo, video, map, and finger-drawing in every note in this folder.

IN THIS CHAPTER

» Setting up your email accounts

» Reading and managing email messages

» Sending email messages

Chapter 10 The Email Must Get Through

Your iPhone's built-in Mail app can send and receive both text email messages and fully formatted messages including graphics. The Mail app works with most popular email providers with a minimal amount of setup and with other email providers with just a little more work.

Mail enables you to use a unified inbox that gathers messages from all the email accounts you have set up. You can organize messages by *thread,* or conversation, for quick and easy reading. You can mark important senders as VIPs for special treatment. And through the Continuity and Handoff features, you can start writing an email on your iPhone, continue it on your iPad, and finish it on your Mac.

Prep Work: Setting Up Your Accounts

To use Mail, you need to add your email accounts to your iPhone. If you signed in to iCloud when setting up your iPhone, iOS automatically added your iCloud account to the iPhone, and your iCloud email account should already be set up and functional, unless you've disabled it. You can add other accounts as needed.

Checking that your iCloud account is set up for Mail

To make sure your iCloud account is set up for email, tap the Mail app icon on the Home screen. If the Mailboxes screen appears, showing your iCloud account, you're all set.

If the Mailboxes screen doesn't show your iCloud account but shows another account, or if the Welcome to Mail screen appears instead of the Mailboxes screen, you'll need to enable mail for your iCloud account. Choose Settings ➪ Apple ID ➪ iCloud ➪ iCloud Mail, and then set the Use on This iPhone switch on (green).

Adding other email accounts

To add another email account, choose Settings ➪ Mail ➪ Accounts ➪ Add Account. The Add Account screen appears (see Figure 10-1). Continue with the next subsection to add an account on iCloud, Gmail, Yahoo!, AOL, or Outlook.com. Continue with the "Setting up corporate email" subsection if you're setting up company email through Microsoft Exchange. Continue with the subsection "Setting up an account with another provider" for a provider whose name doesn't appear.

FIGURE 10-1: On the Add Account screen, tap the provider of the account you want to add.

If you need to get a free email account, look at Yahoo! (`https://mail.yahoo.com/`), Google (`https://mail.google.com/`), Microsoft (`www.outlook.com`), AOL (`www.aol.com`), or one of many other service providers.

Setting up an email account on iCloud, Gmail, Yahoo!, AOL, or Outlook

If your account is with Apple's own iCloud service, Google's Gmail, Yahoo!, AOL, or Microsoft's Outlook.com, tap the appropriate button on the Add Account screen, and then follow through the resulting screens. These screens vary depending on the email provider, but you'll always need to provide your email address and your password. You'll usually also need to go through two-factor authentication or secondary verification. For example, you may need to enter a security code that the provider texts to you.

When iOS displays a screen for choosing which account features to use (see the Gmail example in Figure 10-2), set each switch on (green) or off (white), as needed.

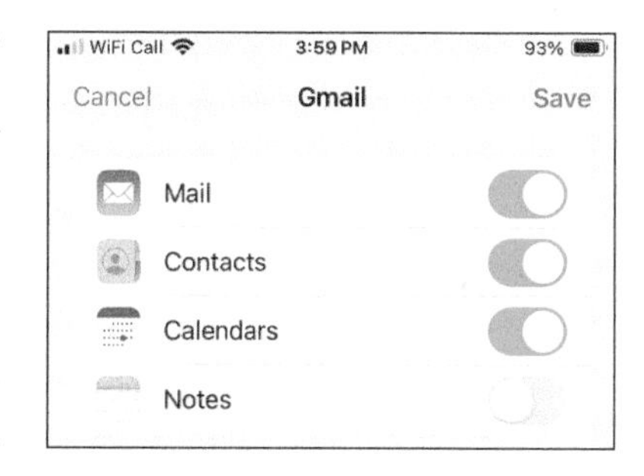

FIGURE 10-2: Set the switches to choose which account features to use on your iPhone.

Setting up an account with another provider

If your email account is with a provider other than iCloud, Exchange, Gmail, Yahoo!, AOL, or `Outlook.com`, make sure you have the following information:

- Your email address and password
- The hostname for the incoming mail server (such as `mail.providername.com`)
- The incoming mail server type, either Internet Mail Access Protocol (IMAP) or Post Office Protocol (POP, also called POP3)
- The hostname for the outgoing mail server (such as `smtp.providername.com`)
- Whether you need to provide your user name and password for the outgoing mail server

Armed with this information, set up your account like this:

1. **On the Add Account screen, tap the Other button.**
2. **On the second Add Account screen, tap Add Mail Account to display the New Account screen.**
3. **In the Name box, type your name the way you want it to appear on outgoing messages.**
4. **In the Email box, type the account's email address.**
5. **In the Password box, type the account's password.**
6. **In the Description box, type a description of the account (to help you identify it).**

7. **Tap the Next button.**

   Your iPhone attempts to validate the email account using the information you've given. If it succeeds in doing so, skip ahead to Step 12. If not, continue with the following steps.

8. **Tap the IMAP tab button or the POP tab button to specify the incoming mail server type.**

9. **In the Incoming Mail Server section, enter the server's host name, the account's user name, and the account's password.**

10. **In the Outgoing Mail Server section, enter the server's host name, the account's user name, and the account's password.**

11. **Tap the Next button.**

    iOS contacts the servers and verifies your credentials. The IMAP screen or the POP screen then appears so that you can choose which account features to use. This screen is like the Gmail screen shown in Figure 10-3 but may contain fewer switches, such as only a Mail switch and a Notes switch.

12. **Set the switch on (green) for each feature you want to use.**

13. **Tap the Save button to save the account.**

    The account then appears on the Accounts screen in Mail settings.

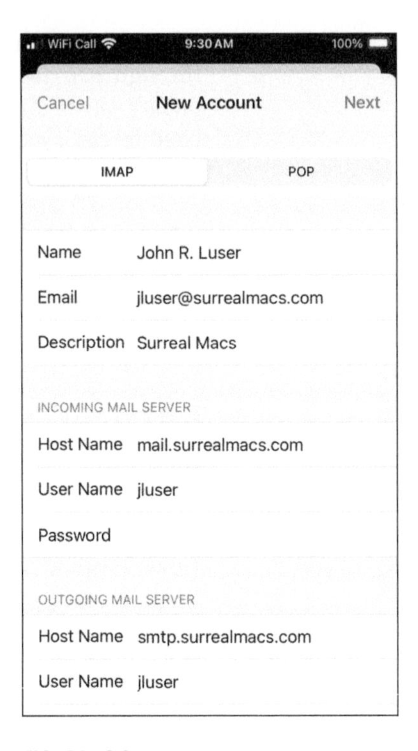

FIGURE 10-3: You may need to enter mail server details when adding an email account.

Setting up corporate email

If your company or organization uses Microsoft's Exchange Server technology to manage its email, calendaring, and scheduling, you can connect your iPhone to the Exchange Server system by adding your account to it. The Microsoft Exchange ActiveSync feature can *push* (send) updates to your iPhone automatically, ensuring that you get them immediately rather than when your iPhone checks for new mail and other data.

WARNING

Connecting your iPhone to an Exchange Server system typically enables the Exchange administrator to remotely erase your iPhone without warning. For this reason, think hard before connecting your own iPhone to an Exchange Server system when your employer offers a bring-your-own-device (BYOD) arrangement: Is the convenience of having your Exchange email and other data (contacts, calendar, reminders, notes) to hand worth the risk of having your iPhone wiped without warning? Using a corporate device is much safer for you, even if it is inconvenient.

To set up an Exchange account, follow these steps:

1. **Choose Settings ➪ Mail ➪ Accounts ➪ Add Account to display the Add Account screen (refer to Figure 10-1).**
2. **Tap the Microsoft Exchange button to display the Exchange screen.**
3. **In the Email box, type the email address for your Exchange account.**
4. **In the Description box, type a description (such as Work Email).**
5. **Tap the Next button.**

   iOS looks up Exchange servers that know your email address.
6. **In the Sign In to Your Exchange Account Using Microsoft? dialog, tap the Sign In button, and then authenticate yourself with your password.**
7. **On the screen shown in Figure 10-4, set the Mail switch on (green). Set the Contacts switch, Calendars switch, Reminders switch, and Notes switch on (green) or off (white), as appropriate.**

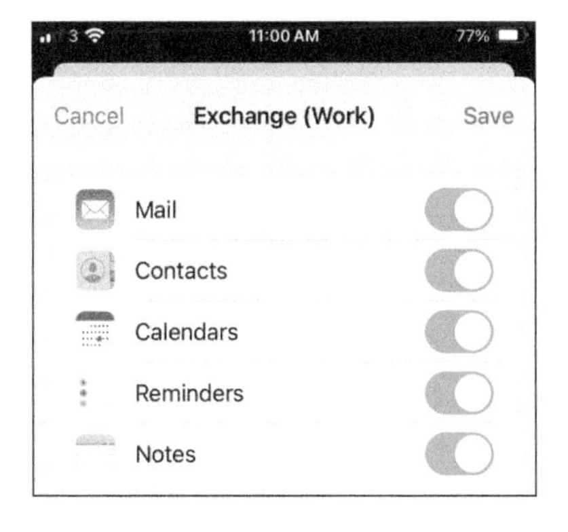

FIGURE 10-4: On this screen, enable each Exchange feature you want to use.

8. **Tap the Save button to finish adding your Exchange account.**

   The Exchange account appears on the Accounts screen.

TIP

While you're on the Accounts screen, tap the button for your Exchange account (the button shows the descriptive name you entered) to display the configuration screen for the account. Tap the Mail Days to Sync button, and then tap No Limit, 1 Day, 3 Days, 1 Week, 2 Weeks, or 1 Month, as needed.

See Me, Read Me, File Me, Delete Me: Working with Messages

You can tell when you have unread mail by looking at the Mail icon on your Home screen. The total number of unread messages across all email inboxes appears in the red badge on top of the icon.

Reading messages

Tap the Mail icon to display the Mailboxes screen (see Figure 10-5). If you've enabled Mail for more than one account, the All Inboxes inbox appears at the top, giving you unified access to all the accounts. The number to the right of All Inboxes shows the total number of unread messages in all your accounts.

Below the All Inboxes inbox are the inboxes for your individual email accounts, each showing how many unread messages it contains.

Further down the Mailboxes screen is a list of your email accounts in a larger, bold font. Tap an account to expand it, showing the subfolders it contains, such as Drafts, Sent, and Junk (see Figure 10-6). Tap the account name again to collapse the account.

From the Mailboxes screen, you can start creating a mailbox by first tapping the Edit button in the upper-right corner to enable edit mode, and then tapping the New Mailbox button in the lower-right corner. On the New Mailbox screen, type the name for the mailbox; tap the Mailbox Location button, and then tap the account on the Mailbox Location screen (and, optionally, the existing mailbox within which to create the new mailbox); and then tap the Save button. Back on the Mailboxes screen, tap the Done button when you finish adding mailboxes.

To read your new mail, either tap All Inboxes to see all your inbox messages in a unified view or tap an individual account's inbox to see only that account's messages. To make sure you've got the latest messages, refresh the view by swiping down a short way on the middle of the screen. Mail displays a spinning gear icon briefly while it checks for new messages.

When you open a mailbox, Mail fetches the most recent messages. Tap a message to read it. When a message is on the screen, icons for managing that message appear below it; see the next subsection for details.

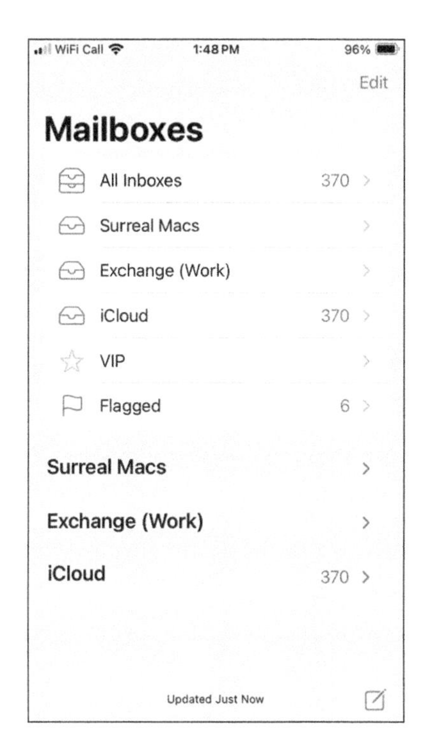

FIGURE 10-5:
The Mailboxes screen is divided by inboxes and accounts.

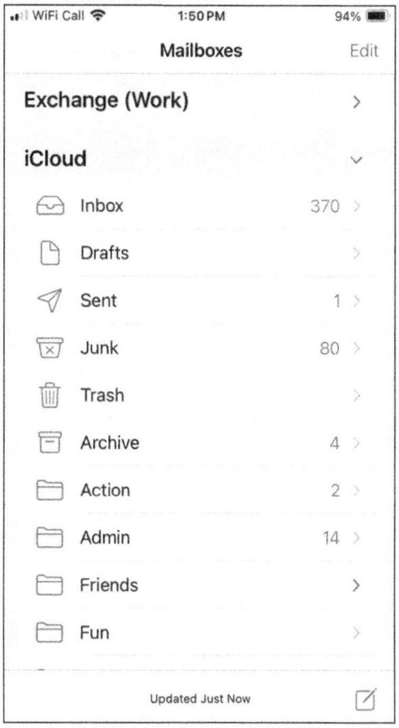

FIGURE 10-6:
Tap one of your email accounts to reveal its subfolders.

In landscape mode on larger-screen iPhones, you can get a more expansive view that takes advantage of the extra display real estate. In the left pane, you see the header and first line from messages in your inbox. In the right pane, you see part of the contents of the highlighted email.

Filtering mail

When a mailbox gets full of messages, you can filter it to make it display only the messages you're interested in. Tap the filter icon (shown in the margin) to enable filtering; the icon reverses its colors, making it mostly blue, and the Filtered By readout at the bottom of the screen shows the current filter applied. In the left screen in Figure 10-7, the current filter is Unread.

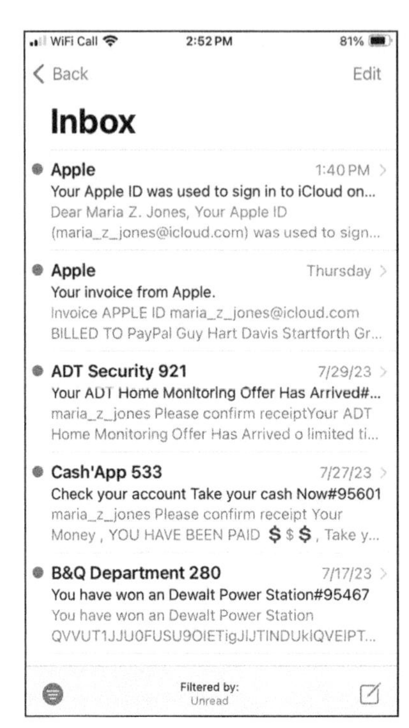

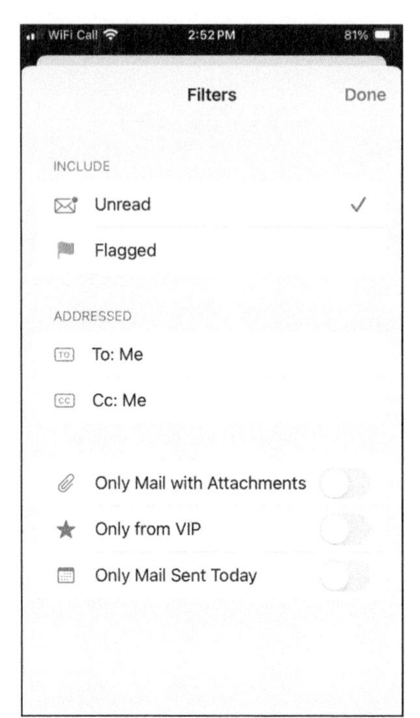

FIGURE 10-7: You can easily determine how to filter the emails in your inboxes.

To apply a different filter, tap the current filter (here, Unread) to display the Filters screen, shown on the right in Figure 10-7. Make your choices in the Include box and the Addressed box, and set the Only Mail with Attachments switch, the Only from VIP switch, and the Only Mail Sent Today switch on (green) or off (white), as needed. Then tap the Done button to apply those filters and return to your mailbox, whose contents now reflect the choices you made. When you want to see all your messages again, tap the filter icon again to remove the filtering.

Managing messages

When a message is on your screen, you can do many tasks in addition to reading it. Check out Figure 10-8 for the location of the controls mentioned in this section.

By tapping the icons labeled in Figure 10-8, left, or the menu choices shown in Figure 10-8, right (which you can get to by tapping the more actions icon), you can perform the following actions:

- View the next message.
- View the preceding message.
- Flag the message as important.

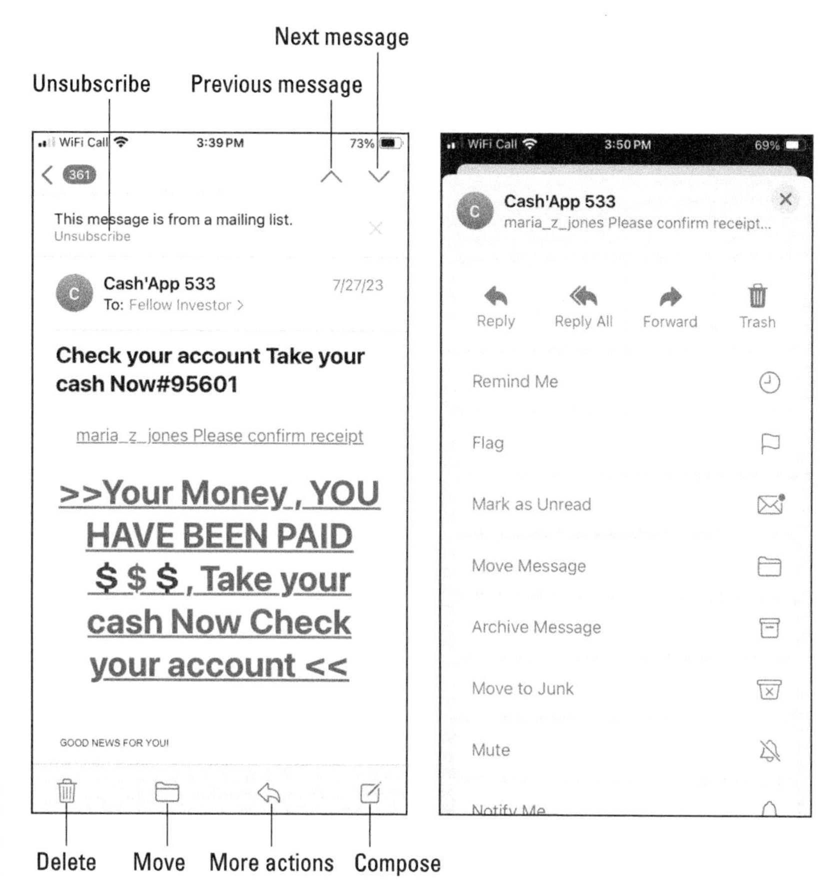

FIGURE 10-8: Reading and managing an email message.

- Move the message to the Junk folder.
- Archive the message.
- Mark the message as unread (or, if it is unread, mark it as read).
- Get notified when anyone replies to the email thread.
- Mute or unmute notifications from the email thread.
- Move the message to another folder. When the list of folders appears, tap the folder you want to move the message to.
- Delete the message.

If you tap the delete icon (trashcan) by mistake, you'll need to dig the message out of the trash. If you'd like Mail to double-check before deleting a message, choose Settings ⇨ Mail, go to the Messages box, and then set the Ask Before Deleting switch on (green).

- Reply to the sender or to the sender and all recipients.
- Forward the message to someone else.
- Print the message.
- Create an email message.

You can delete email messages in several ways without opening them:

- Swipe left a short distance across the message's preview in the mailbox, so that the gray More button, the orange Flag button, and the red Trash button appear. Tap the red Trash button.
- Swipe left all the way across the message's preview in a mailbox, so that the red Trash button swallows up the whole preview.

REMEMBER

Google's Gmail is set to archive messages by default rather than delete them. If the purple Archive button appears instead of the red Trash button when you swipe left on a message, you'll know the account is set to archive messages. If you want it to delete messages instead, choose Settings ➪ Mail ➪ Accounts ➪ Gmail, tap the account name, and then tap Advanced on the Account screen. Go to the Move Discarded Messages Into box, and then tap the Deleted Mailbox button, moving the check mark to that button from the Archive Mailbox button.

- Tap the Edit button (in the upper-right corner of the Inboxes screen or a mail folder screen), and then tap the selection circle to the left of each message you want to remove, as shown in Figure 10-9. Tapping that circle puts a check mark in it. Tap the Trash button in the lower-right corner of the screen to delete all the messages you selected. Deleted messages are moved to the Trash folder.
- Trash the message (or mark it as read) inside an interactive notification that appears regardless of the screen you're viewing. The advantage to this method is you don't have to leave the screen to delete the message. Just tap the Trash button in the notification.

TIP

- To move multiple messages at once, tap Edit, and then tap the selection circle to the left of each message, putting a check mark in it. Tap the Move button at the bottom of the screen (see Figure 10-9), and then tap the folder to which you want to move those messages.
- When you've opened a message to read it, you can drag from the very left edge of the screen to the right to peek into the mailbox that contains the message (see Figure 10-10).
- Drag your finger to the left across a message to display the More button, the Flag button, and the Trash button. Drag your finger to the right across a message to display a button for toggling the message's status between read and unread.

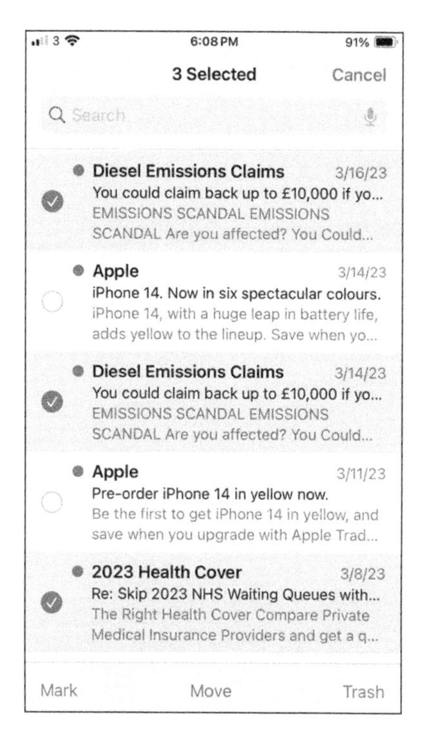

FIGURE 10-9:
Select multiple messages to move them or delete them quickly.

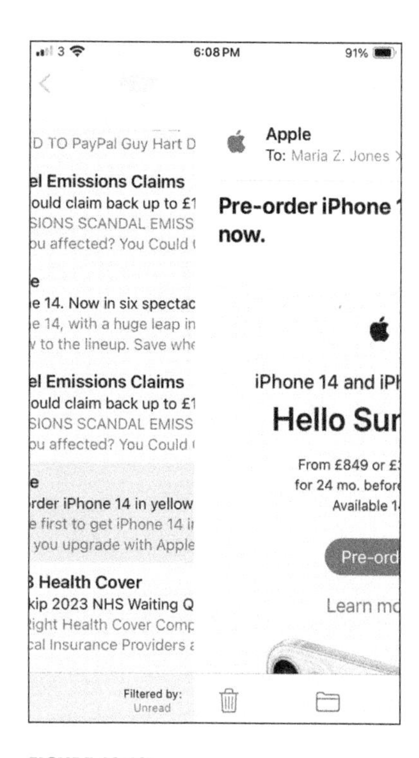

FIGURE 10-10:
Drag the very left edge of your message to the right to peek at the mailbox that contains the message.

You can also take the following actions by swiping across a message's preview in a mailbox:

* **Flag the message:** Swipe left partway across the preview, and then tap the orange Flag button.
* **Mark the message as read or unread:** Swipe right partway across the preview, and then tap the blue Read button (if the message is unread) or the blue Unread button (if the message has been read).

TIP

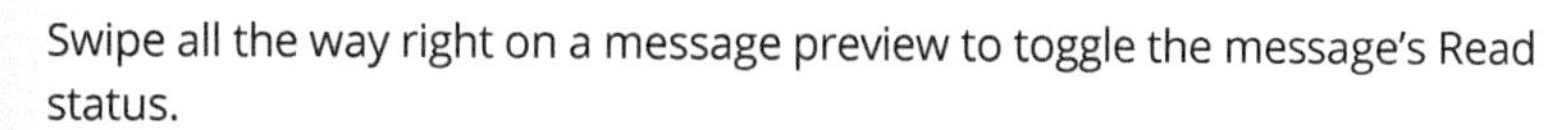

Swipe all the way right on a message preview to toggle the message's Read status.

* **Set a reminder on the message:** Swipe right partway across the preview, tap the purple Remind Me button, and then tap the appropriate item on the Remind Me menu, such as the Remind Me in 1 Hour item or the Remind Me Tomorrow item. To set a specific time for the reminder, tap the Remind Me Later item, set the date and time on the Remind Me screen, and then tap the Done button.

Threading messages

Mail enables you to display your messages in *threads,* also called *conversations,* so that you can easily follow a conversation.

When you organize messages by thread or conversation, the related messages show up as a single entry in the mailbox, with a circled, blue, right-pointing arrow indicating a thread. If a message is not part of a thread, you see a single right-pointing gray arrow. In the left screen in Figure 10-11, you can see the thread arrow on the message. Tap the thread arrow to display the messages in the thread (see the right screen in Figure 10-11).

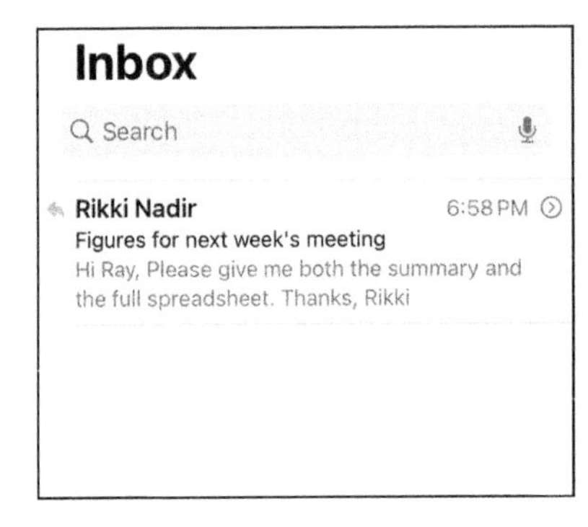

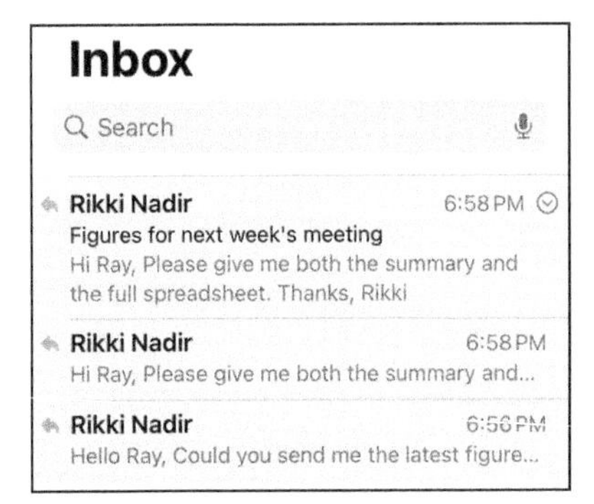

FIGURE 10-11: Threaded messages appear as a single entry in the mailbox (left). Tap the thread icon to expand the thread (right).

To turn on threading, choose Settings ⇨ Mail, go to the Threading section, and then set the Organize by Thread switch on (green). You can also choose several other settings in the Threading section:

- **Collapse Read Messages:** Set this switch on (green) to collapse thread messages you have read.
- **Most Recent Messages on Tap:** Set this switch on (green) to put the most recent messages at the top of the thread. This feature is usually helpful.
- **Complete Threads:** Set this switch on (green) to have the thread show all the messages, including those in different mailboxes.
- **Muted Thread Action:** Tap this button, and then tap the Mark as Read button or the Archive or Delete button to specify how to handle threads you've muted.
- **Blocked Sender Options:** Tap this button, and then set the Mark Blocked Sender switch on (green) to mark blocked senders. In the Actions box, tap the Leave in Inbox button or the Move to Trash button, as appropriate.

Searching emails

To search for particular messages, tap the search field at the top of an inbox or a mailbox, and then start typing your search term. Mail displays matching results as you type. When you finish typing the search term, you can tap the Search button on the keyboard to hide the keyboard, giving you more space to examine the results.

Tap the All Mailboxes button below the search field search all your mailboxes, or tap the Current Mailbox button to search only the current mailbox.

The Top Hits section presents the most promising search results. Below that, the Suggestions section suggests more specific searches, such as by people, by attachments, or by message subject. Tap a suggestion to perform that search.

TIP

You might want to create a shortcut for searches you perform regularly. Choose Settings ⇨ Mail, and then tap the Siri & Search button to display the Siri & Search screen. Set the Learn from This App switch on (green) to tell Search to learn from Mail. In the While Searching box, set the Show App in Search switch and the Show Content in Search switch on (green).

Dealing with attachments

Attaching files to an email message can be a great way of transferring files from one person or device to another. Mail makes it easy to work with files attached to incoming messages.

How an attached file appears depends on whether the file uses a file type that iOS supports natively. iOS supports a good variety of file types natively, including the leading image formats (JPEG, TIF, GIF, PNG, and HEIC); standard document formats, such as Word documents, Excel workbooks, and Adobe's Portable Document Format (PDF); and audio files (such as the AAC, MP3, and Apple Lossless Encoding formats).

Mail typically displays images and videos within the body of the email, so that they appear as part of the email text. Mail places most other file types at the end of the message, as you see in the left screen in Figure 10-12. Here, you can tap the download icon (shown in the margin) to display the pop-up menu, and then tap Save to Documents, Save to Downloads, or Save Attachment (to the folder you specify). Alternatively, tap the attached file's name or icon to open it for viewing (see the right screen in Figure 10-12). From here, you can tap the share button in the lower-left corner to use the file in other ways (such as opening it in the iOS Pages app, in this instance) or simply tap the Done button to return to the email message.

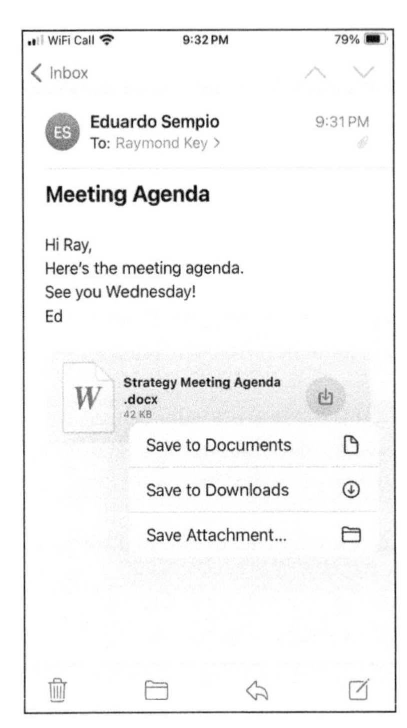

FIGURE 10-12: Tap the Download icon to save an attached file (left). Tap the file's name or icon to view the file in Mail (right).

More things you can do with messages

Here are ten more moves you'll want to use on your incoming messages:

- **See details for recipients.** Tap the header area of the message, and then tap a recipient's name (displayed in blue). This works for all To recipients and Cc recipients. You don't get to see Bcc recipients.
- **Add a sender or recipient to your contacts.** Tap the header area, tap the sender or recipient's name, and then tap either Create New Contact or Add to Existing Contact.

- **Mark a message as read or unread.** Tap the more actions icon (shown in the margin), and then tap the Mark as Read button or the Mark as Unread button, as appropriate.
- **Flag a message.** Tap the more actions icon, and then tap the Flag button. To change the flag color, tap a color on the bar below the Unflag button (which replaces the Flag button).
- **Mark a message as junk.** Tap the more actions icon, and then tap the Tap Move to Junk button.
- **Tell Mail to notify you about thread updates.** Tap the more actions icon, and then tap the Notify Me button.

» **Silence a conversation.** Tap the more actions icon, and then tap the Mute button.

» **Make a sender a VIP.** Tap the header area, tap the name or email address at the top of the message, and then tap Add to VIP. Mail displays a gold star next to each message from a VIP, helping you spot them easily. You can also view mail from all your VIPs by tapping the VIP folder in the list of mailboxes.

TIP

To pull the red carpet out from under a VIP, tap the header area, tap the VIP's name, and then tap Remove from VIP.

» **Zoom in or out on a message.** Unpinch to zoom in, pinch to zoom out.

» **Follow a link in a message.** To follow a link in a message, tap the link. (Links are typically displayed in blue and may be underlined, but sometimes they appear in other colors.) If the link is a URL, Safari opens and displays the web page. If the link is a phone number, the Phone offers to dial the number. If the link is an address, Maps opens. If the link is a day, date, or time, you can tap the item to create a calendar event. If the link is a shipper's tracking number, you may be able to get the status of a package. If the link is an email address, Mail creates a pre-addressed blank email message.

TIP

If tapping the link opens a different app, tap the tiny Mail button in the upper-left corner when you want to return to Mail.

WARNING

Never tap a link or attachment if you can't verify or trust the sender of an email. You're opening yourself up to a potentially serious security risk.

Darling, You Send Me (Email)

The Mail app enables you to send text-only messages, messages that include photos or videos, and messages that have files attached. You can also save unfinished messages as drafts, reply to messages you've received, or forward messages to others.

Sending an email message

To compose a new email message, follow these steps in the Mail app:

1. **Tap the compose icon (shown in the margin) in the lower-right corner of the screen.**

   The New Message screen appears (see Figure 10-13).

2. **Enter the names or email addresses of the recipients in one of the following ways:**

   - *Type the names or email addresses of the recipients in the To field.*
   - *Tap the microphone key to dictate the names or email addresses of the recipients in the To field.*
   - *Tap the + icon to the right of the To field to select a contact or contacts from your iPhone's Contacts app.*

TIP

You can rearrange names in the address fields by dragging them around. For example, you can drag a name from the To field to the Cc field.

Mail may suggest adding recipients you have used for previous group messages that include the people you've already added.

FIGURE 10-13: The New Message screen appears, ready for you to start typing the recipient's name.

3. **(Optional) Enter a name in the Cc field, Bcc field, or both fields. Or choose to send mail from a different account in the From field, as follows:**

   (a) *Tap the field labeled Cc/Bcc, From.*

   Doing so expands the field to show the Cc field, the Bcc field, and the From field separately. The Cc label stands for *carbon copy,* and *Bcc* stands for *blind carbon copy.* Bcc enables you to include a recipient on the message that other recipients can't see has been included.

TIP

   Bcc is great for messages to groups of people who may not know each other and who don't need each other's email addresses. For example, if you send out a change-of-address email to everyone you know, put your email address in the To field and all the other addresses in the Bcc field. That way, each other recipient sees only your address and their own address.

   (b) *Tap the respective Cc or Bcc field and type the name.*

   Or tap the + icon that appears in one of those fields to add a contact.

TIP

   If you start typing an email address, email addresses that match what you typed appear in a list. If the correct one is in the list, tap to use it.

(c) Tap the From field if you need to change the address that's sending the message.

The current email account has a check mark next to it. Tap another email address to send the message from that account, or tap Hide My Email to create a random address that forwards to your inbox.

4. **In the Subject field, type or dictate a subject.**

   The subject is optional, but it helps the recipient decide whether (or when) to read the message. A missing subject may also make your message look like junk mail.

5. **In the message area, type or dictate your message.**

   The message area is immediately below the Subject field.

   Turn your iPhone to landscape orientation if you want to use the wider-format keyboard. This makes typing easier but reduces the amount of text you can see.

   If you type or paste a text link in the message, you can convert it to a graphical link by tapping the link, tapping the downward-pointing arrow that appears, and then tapping Show Link Preview.

6. **Tap the Send button (the up arrow) in the upper-right corner of the screen.**

Provided your iPhone has an internet connection, Mail sends the message almost immediately — but not quite immediately. Instead, Mail implements a short delay to give you a chance to undo sending the message.

If you realize there's something wrong with the message, tap the Undo Send button at the bottom of the screen. By default, Mail gives you 10 seconds to undo sending.

TIP

To increase the time to undo sending, choose Settings ➪ Mail ➪ Undo Send Delay, and then tap the 20 Seconds button or the 30 Seconds button. You can also tap the Off button to disable the Undo Send Delay feature.

TIP

Instead of sending the message immediately, you can schedule it for sending later. To schedule the message, long-press the Send button until the pop-up menu opens. Then either tap the suggested time (such as Send 8:00 AM Tomorrow) or tap Send Later and specify the date and time you want.

REMEMBER

Mail will prompt you to follow up on a message you sent with such phrases as *Could you send* or *I will do by* that never got a response. You will also receive a prompt if the app detects that an attachment or recipient is missing before you tap Send.

Sending a photo or video with an email

To add a photo or video to a message you've already started, follow these steps:

1. **Long-press the place in the message where you want to add the photo or video, and then lift your finger.**

   The pop-up command bar appears, showing buttons such as Select and Select All.

2. **Tap the > button one or more times until the Insert Photo or Video button appears.**
3. **Tap the Insert Photo or Video button to display the photo browser at the bottom of the screen.**
4. **Tap either Recent Photos or All Photos, and then browse to the photo or video.**
5. **Tap each photo or video, placing a check mark on it.**
6. **Tap the *X*-in-a-circle button to close the photo browser and insert the photos or videos in the message.**

   If you selected a video, Mail compresses it and inserts a thumbnail of it in the body of the message. The recipient can play the video by tapping the thumbnail.

WARNING

If the video is too long or large, you may not be able to send it by email — internet service providers (ISPs) and email providers typically impose limits. However, if your outgoing message and video or other attachments total 5GB or less, you have the option of sending them over iCloud via Mail Drop. Your recipient will have 30 days to download a Mail Drop payload.

If you choose a photo, you'll see the picture inserted in the message body. When you're ready to send your email, tap the Send button. If Mail displays a screen prompting you to choose which photo size to send (see Figure 10-14), tap the appropriate button.

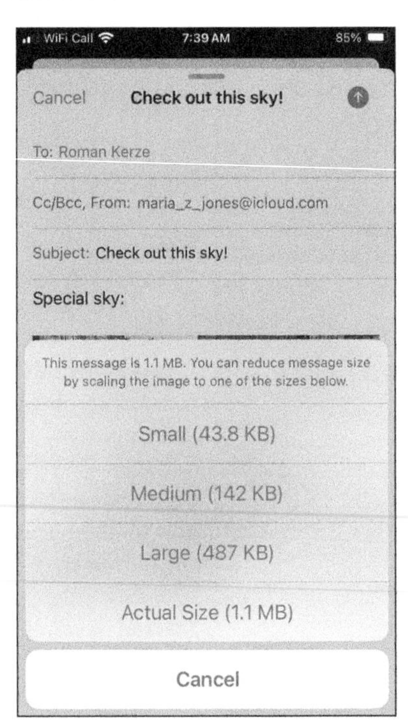

FIGURE 10-14: You may need to tap a button to specify which size of photo to send.

Another way to include a photo or video in an email is to tap the Photos icon on the Home screen, and then find the photo or video you want to send. Tap the share icon (shown in the margin), and then tap the Mail button. Mail creates a new message containing the photo or video. You can then address the message, add a subject and any text needed, and send it.

Saving an email message as a draft

Sometimes you start an email message but don't have time to finish it. When that happens, you can save it as a draft and finish it some other time. Tap the Cancel button in the upper-left corner of the screen, and then tap the Save Draft button in the dialog that appears. Mail saves the message in the Drafts folder.

WARNING

The dialog also contains the Delete Draft button, which gets rid of the message instantly. Make sure you don't tap this button by mistake.

TIP

To work on the message again, long-press the compose icon until the Drafts screen appears, and then tap the message to open it. If you prefer, you can open the Drafts mailbox by tapping it on the Mailboxes screen, and then tap the message.

You can also move a draft message out of the way temporarily so that you can look at or work on other messages. From the top of the New Message screen, drag down the draft of the message you're composing until it becomes a button at the bottom of the screen (see the left screen in Figure 10-15, which shows a button for the Contribution for Report message). You can then work with other messages or mailboxes as usual, ignoring the button at the bottom of the screen. You can park multiple messages at the bottom of the screen like this; only the last message's name appears.

When you're ready to resume work on one of the parked messages, tap the button at the bottom of the screen. If there's only one parked message, it appears full screen. If there are multiple parked messages, Mail displays previews so that you can tap the message you want (see the right screen in Figure 10-15).

Formatting text in an email

Mail lets format email text by applying bold, italic, or underline to it. First, select the text by long-pressing a word until the pop-up bar appears (see the left screen in Figure 10-16); drag the selection handles to adjust the selection as needed. Tap the > button until you see the Format button (see the center screen in Figure 10-16). Tap the Format button, and then tap the Bold button, the Italic button, or the Underline button (see the right screen in Figure 10-16).

If you tap Quote Level — another option that appears when you tap the > button after selecting a word — you can quote a portion of a message you're responding to. You can also increase or decrease the indentation in your outgoing message.

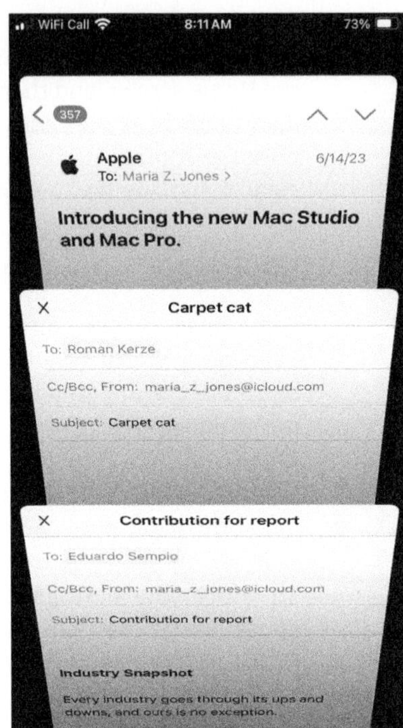

FIGURE 10-15: You can park one or more open messages at the bottom of the screen (left). Tap the button to display thumbnails of the messages (right), and then tap the message you want.

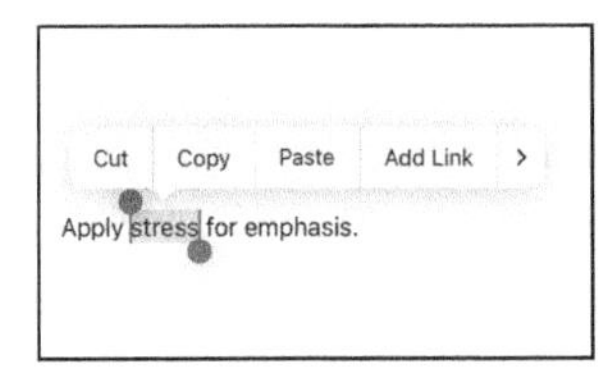

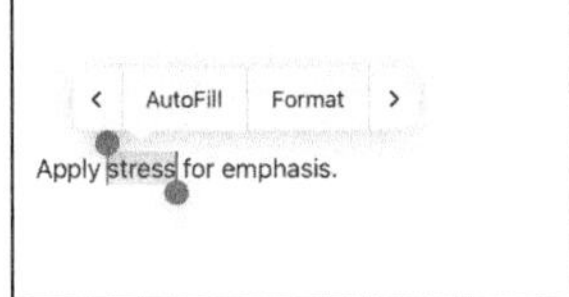

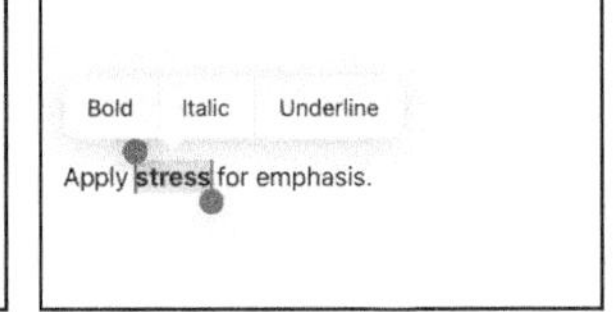

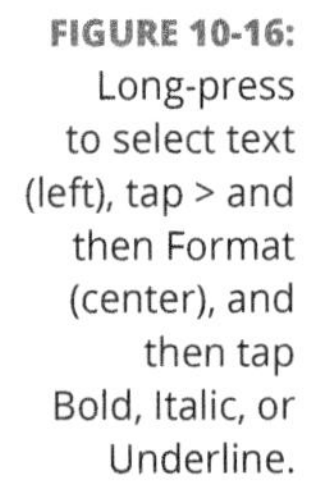

FIGURE 10-16: Long-press to select text (left), tap > and then Format (center), and then tap Bold, Italic, or Underline.

If you tap Replace, you'll see other word options for replacing the word you selected, perhaps *formal* instead of *format*. Tap Look Up, and you can summon a dictionary definition for the highlighted word and check out relevant web searches and personalized suggestions from iTunes and the App Store. If appropriate, you may also see movie showtimes or nearby locations.

You'll also see a Share button. Tap it and you are brought to the same options presented when you tap the share icon at the bottom of the screen.

Using the formatting and attachments toolbar

Instead of applying formatting as explained in the preceding subsection, you can apply formatting using the formatting and attachments toolbar (shown on the left in Figure 10-17). To display this toolbar, long-press a word. If you just see the predictions bar suggesting replacement words, tap the < button on the right of this bar.

Tap the text format icon (Aa) to display the Format pane, shown on the right in Figure 10-17. Here, you can apply bold, italic, underline, and strikethrough; change the font and font size; create numbered or bulleted lists; and change the text alignment, indentation, and line spacing.

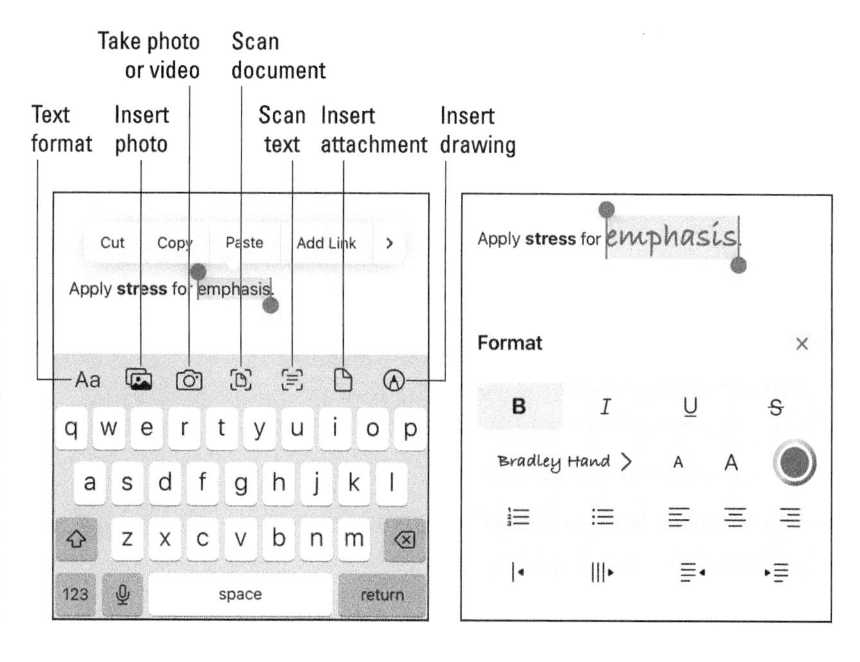

FIGURE 10-17: Tap the Aa button on the formatting and attachments toolbar (left) to display the Format pane, which lets you change the text font, size, color, and more (right).

Adding Live Text

When you want to insert some text from a hard-copy document, position the insertion point at the appropriate point in the message and then tap the Scan Text button on the formatting and attachments toolbar to activate the Live Text feature. The bottom part of the screen displays a viewfinder from the iPhone's rear camera. Aim the camera at the text you want to capture, so that yellow brackets appear around the text, and then tap the Insert button.

Adding attachments

To add an attachment to an outgoing message, tap the Insert Attachment button on the formatting and attachments toolbar. In the Files app overlay that appears, navigate to the folder that contains the file you want to insert, and then tap the file. Mail adds an icon for the file to the body of the message.

Marking up attachments

If you've inserted a photo or a PDF document in an outgoing message, you can use the Markup feature to mark it up. Tap the photo or PDF to select it, and then tap the Markup button on the formatting and attachments toolbar. You can then draw on the photo or PDF by tapping the simple annotation tools below it.

The tools are represented by drawing icons (see Figure 10-18), including a pen, a highlighter, and an eraser. (Unless you use an optional stylus, your finger will be that pen.) There's also a magnification loupe so that you can zoom in on the photo or PDF, and options to write text either freehand or with the keyboard.

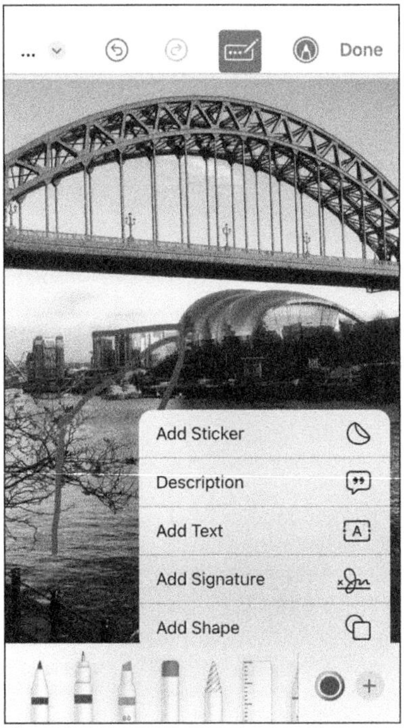

You can draw arrows, circles, and squares. And you can change the color and thickness of the lines and symbols you draw.

Lastly, you can write your signature in script. Tap the circled +, and then tap Add Signature. You can tap an existing signature if you've already created one. Or tap Add or Remove Signature, tap add (+) in the upper-left corner of the screen, and then sign your name with your finger.

Mail also lets you draw inside an email using the Markup tools. Without anything selected in the message body, tap the Markup button on the formatting and attachments toolbar. You can then draw using the drawing tools. Tap Done to insert the drawing into the message.

FIGURE 10-18: Simple tools let you mark up a photo or PDF attachment before sending it.

Replying to or forwarding an email message

When you receive a message and want to reply to it, open the message, and then tap the more actions icon (shown in the margin). Then tap the Reply button, the Reply All button, or the Forward button, as appropriate.

The Reply button creates a new email message addressed to the sender of the original message. The Reply All button creates an outgoing email message addressed to the sender and all other recipients of the original message, except Bcc recipients. In both cases, the Subject line is retained with a *Re:* prefix added. So if the original

Subject line were *iPhone Tips*, the reply's Subject line would be *Re: iPhone Tips.* You also see text from the original message in the body of your reply (whether you are replying to one person or more than one person).

Tapping the Forward button creates an unaddressed email message that contains the text of the original message. Add the email address of each person you want to forward the message to. When you forward a message, Mail adds the *Fwd:* prefix to the Subject line rather than the *Re:* prefix, making the sample message's Subject line read *Fwd: iPhone Tips.* If the message you're forwarding contains attachments, Mail lets you choose whether to include them in the forwarded message.

You can edit the Subject line of a reply or a forwarded message or edit the body text of a forwarded message the same way you would edit any other text. It's usually best to leave the Subject line alone (with the *Re:* or *Fwd:* prefix intact) to enable threading.

To send your reply or forwarded message, tap the Send button as usual.

In Mail Settings, you can decide to include attachments with replies when adding recipients. Choose Settings ➪ Mail ➪ Include Attachments with Replies, and then tap Never, When Adding Recipients, Ask, or Always, as needed.

Settings for sending email

iOS and the Mail app enable you to customize the sending and receiving of mail in many ways. The following list tells you about six of the most useful ways:

- **Hear a sound when Mail sends a message.** Choose Settings ➪ Sounds & Haptics ➪ Sent Mail, and then tap the sound you want.
- **Add a signature to each outgoing message.** Choose Settings ➪ Mail ➪ Signature to display the Signature screen. In the box at the top, tap All Accounts if you want to use the same signature for all accounts; tap Per Account to use different signatures (such as for home and for work). If you chose All Accounts, tap in the single box; if you chose Per Account, tap in the box for the first account to which you want to add a signature. Type or paste the text of the signature. If you want to apply bold, italic, or underline, long-press a word; drag the selection handles to select the text you want to format; tap the Format button on the pop-up bar; and then tap the Bold button, the Italic button, or the Underline button. For Per Account, tap in each other box and create the signatures you want.
- **Bcc yourself on each outgoing message.** Choose Settings ➪ Mail, and then set the Always Bcc Myself switch on (green). Normally, your Sent mailbox collects a copy of each sent message, so there's no need to bcc yourself.

- **Mark addresses outside your email domain.** Choose Settings ⇨ Mail ⇨ Mark Addresses, and then type or paste the domain (such as `@surrealmacs.com`) in the Mark Addresses Not Ending With box. Now whenever you send email to a different domain, the email address appears in the To field in red instead of blue. Marking addresses in this way could help warn you before you dispatch confidential or incriminating mail to people who are not meant to see such messages.
- **Add indentation to replies and forwarded messages.** Choose Settings ⇨ Mail ⇨ Increase Quote Level, and then set the Increase Quote Level switch on (green).
- **Set your default account for sending email.** When you've added multiple email accounts, choose Settings ⇨ Mail ⇨ Default Account, and then tap the account to use as the default. Mail uses this account to send messages started from other apps, such as when you select a photo in the Photos app and choose Share ⇨ Mail.

Choosing settings for checking and viewing email

Here are the key settings to configure for checking and viewing email:

- **Control how Mail checks for new messages.** Choose Settings ⇨ Mail ⇨ Accounts ⇨ Fetch New Data to display the Fetch New Data screen. Set the Push switch at the top on (green). In the box under the Push switch, verify that each email account is set to Push; if not, tap the account to display the account's screen. Here, tap Push if it is available; if not, tap Fetch, unless you prefer to do things the hard way — in which case, tap Manual. Back on the Fetch New Data screen, go to the Fetch box and tap Automatically, Every 15 Minutes, Every 30 Minutes, Hourly, or Manually, as needed. Automatically is usually the best choice.

  The Push method lets the server push messages to your iPhone as soon as they arrive on the server, so you get your messages faster. Use Push if it's available. If your mail provider doesn't support Push, you need to use Fetch, which involves the Mail app checking with the server to get any messages that have arrived. If your iPhone goes into low-power mode, iOS disables both Push and Fetch.

- **Hear a sound when Mail receives a message.** Choose Settings ⇨ Sounds & Haptics ⇨ New Mail, and then tap the sound you want.
- **Set the number of lines in message previews.** Choose Settings ⇨ Mail ⇨ Preview, and then tap None, I Line, 2 Lines, 3 Lines, 4 Lines, or 5 Lines.

- **Choose whether To and CC labels appear in message lists.** Choose Settings ➪ Mail, and then set the Show To/CC Labels switch on (green) or off (white).
- **Customize left-swipe and right-swipe options.** Choose Settings ➪ Mail ➪ Swipe Options, and then work on the Swipe Options screen.
- **Enable or disable the Ask Before Deleting warning.** Choose Settings ➪ Mail, and then set the Ask Before Deleting switch on (green) or off (white). Confirming deletions may be helpful if your hand shakes or you use your iPhone on transit that shakes you.
- **Enable the Mail Privacy Protection feature.** Choose Settings ➪ Mail ➪ Privacy Protection, and then set the Protect Mail Activity switch on (green). Mail Privacy Protection hides your iPhone's internet address (the IP address) and loads remote content (such as images stored on servers) privately. This helps prevent senders from determining whether you've opened a particular message, let alone from learning your iPhone's IP address or location.

If you subscribe to iCloud+, the paid tier of Apple's iCloud service, use the Hide My Email feature to use disposable email addresses rather than your regular email address for any company or app you don't trust. Choose Settings ➪ Apple ID ➪ iCloud ➪ Hide My Email, and then work on the Hide My Email screen. From the iCloud screen, you can also access the Custom Email Domain feature of iCloud+, which enables you to receive email on a custom domain of your choosing.

- **To archive messages rather than delete them.** Choose Settings ➪ Mail ➪ Accounts, and then tap an email account that presents this option — Gmail and Exchange are two. Next, tap the Account name, and then tap Advanced for Gmail or Advanced Settings for Exchange. In the Move Discarded Message Into, tap either Deleted Mailbox or Archive Mailbox, placing a check mark on the button.

Suspending or deleting an email account

If necessary, you can stop using an email account temporarily or delete it permanently from your iPhone:

- **Stop using an email account.** Tap Settings ➪ Mail ➪ Accounts, and then tap the account name. Tap the Mail switch to turn off mail for the account.

Turning off the account doesn't delete the account. Instead, it hides the account from view and stops it from sending or checking email until you turn it on again.

- **Delete an email account.** Tap Settings ➪ Mail ➪ Accounts. Then tap the account name, and tap Delete Account. You must tap Delete from My iPhone or Delete Account (a second time) to proceed.

Choosing advanced settings for an account

Most accounts also enable you to choose advanced settings. To reach these settings, choose Settings ➪ Mail ➪ Accounts, tap the account name, and then tap the Advanced button.

The selection of advanced settings varies by account but may include the following:

- **Set the stay of execution for deleted messages.** On the Advanced screen, tap the Remove button, and then tap the appropriate choice, such as Never, After One Week, or After One Month.
- **Send signed and encrypted messages.** On the Advanced screen, go to the S/MIME box, and then tap the Sign button to display the Sign screen. Set the Sign switch on (green), go to the Certificates box, and then tap the digital certificate to use. (If you don't have a digital certificate, consult your systems administrator.) Back on the Advanced screen, tap the Encrypt by Default button to reach the Encrypt by Default screen. Set the Encrypt by Default switch on (green), reach into the Certificates box again, and tap the digital certificate to use.
- **Choose whether drafts, sent messages, and deleted messages are stored on your iPhone or on your mail server.** On the Advanced screen, go to the Mailbox Behaviors box. Tap the mailbox you want to affect, such as the Drafts Mailbox, and then tap the folder in which to store it. If you choose to store any or all of these mailboxes on the server, you can't see them unless you have an internet connection (Wi-Fi or cellular). If you choose to store them on your iPhone, they're always available, even if you don't have internet access.

WARNING

Change the following settings only under direct advice from your ISP, email provide, or systems administrator.

- **Reconfigure mail server settings.** Tap Host Name, User Name, or Password in the Incoming Mail Server section or the Outgoing Mail Server section of the Account screen, make your changes, and then tap the Done button.
- **Adjust Use SSL, Authentication, IMAP Path Prefix, or Server Port.** Tap Advanced, tap the appropriate item, and then make the necessary changes.

IN THIS CHAPTER

» Tracking appointments and events with Calendar

» Organizing your tasks with Reminders

» Using the clock as an alarm, a stopwatch, and a timer too

Chapter 11 Managing Your Calendar and Commitments

Your iPhone is a wonderful device for keeping tabs on your schedule and your tasks. Use the Calendar app to note all your appointments and events and share them across your devices. Work in the Reminders app to organize your tasks into different categories, keep track of what you've done and what you haven't, and set reminders based on time or location. And take advantage of the Clock app to see the time anywhere around the world, set alarms and timers, and time events precisely with a stopwatch.

Working with Calendar

To keep on top of your appointments and events, work with Calendar by tapping the Calendar icon on the Home screen. The icon shows today's day and date, such as Fri 14, for quick reference.

Calendar lets you use five views:

- **Year:** Use this view (shown on the left in Figure 11-1) to navigate quickly through years and months. Scroll up and down to reach the month you want to view. Tap the month, and it appears in month view.
- **Month:** Use this view (shown in the center in Figure 11-1) to see the days that contain events, which appear as gray dots. Scroll up and down if needed to reach the day you want to display. Tap the day, and it appears in day view.
- **Day:** Use this view (shown on the right in Figure 11-1) to see your events for a particular day. All-day events and birthdays appear above the timeline for the day, which shows each appointment as a block across the times it spans. Each event appears in the color assigned to the calendar that contains it, so you can see which events belong to which calendars. To display other days, swipe left or right, or tap the day on the bar at the top that shows days and dates.

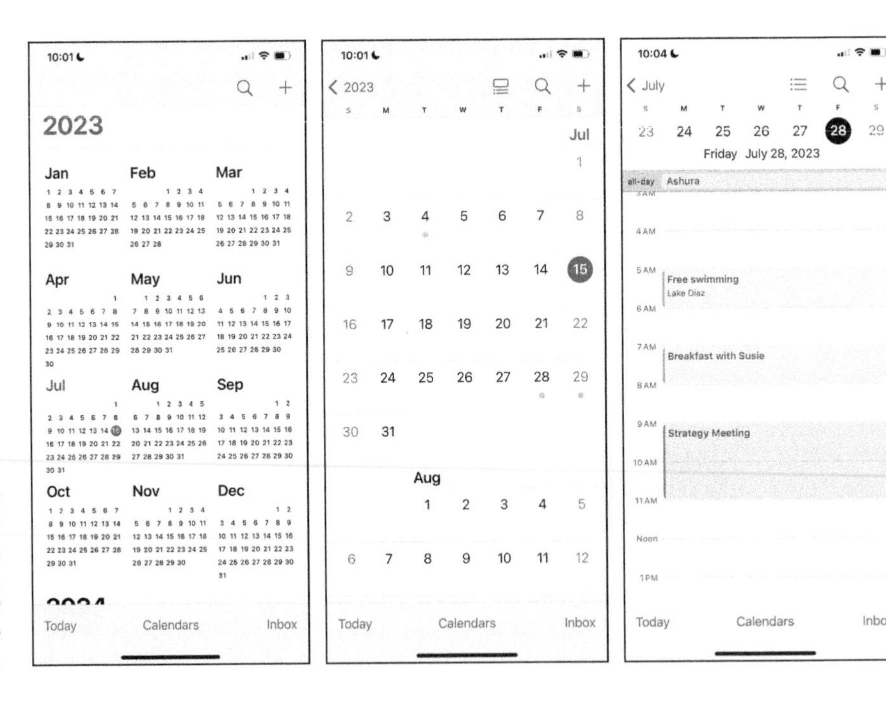

FIGURE 11-1: Year view (left), month view (middle), and day view (right) in the Calendar app.

- **Week:** Turn your iPhone to landscape mode in year view, month view, or day view to switch to week view. Turn the iPhone back to portrait view to return to the previous view.

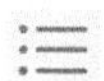

» **List:** In day view, tap the List icon at the top of the screen and shown in the margin to switch to list view. This view (see Figure 11-2) lists your calendar appointments in chronological order. To navigate, you can scroll through the list, or tap the search icon to display the search field, and then type a search term. Tap an event to display its details. Tap the List icon again to switch back to the previous view.

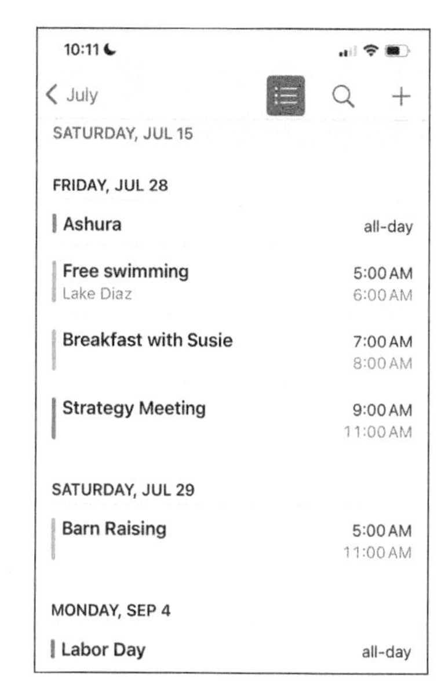

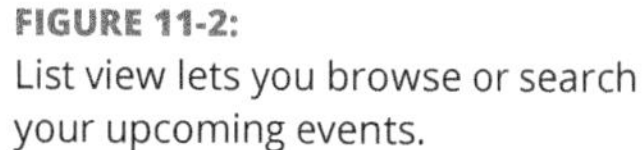
FIGURE 11-2: List view lets you browse or search your upcoming events.

Adding Calendar events

The Calendar app syncs your events from iCloud and from other calendar accounts you add to your iPhone. But you can also create events directly on your iPhone.

TIP

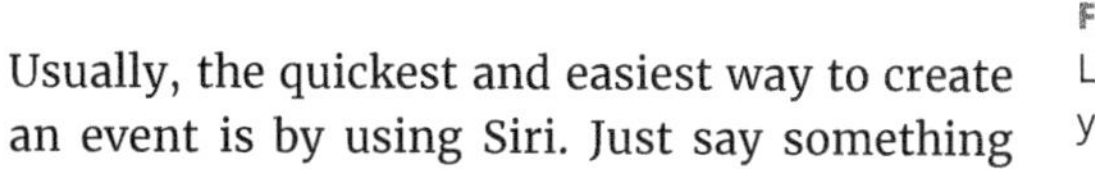
Usually, the quickest and easiest way to create an event is by using Siri. Just say something like "Add a meeting from 10AM to 1PM tomorrow to my Work calendar." If any essential information is missing, Siri prompts you for it. In this case, there's no event name, so Siri asked "What do you want to call it?"

You can also add events manually like this:

1. **In year, month, or day view, tap the add icon (+) in the upper-right corner of the screen.**

   The New Event screen appears (see Figure 11-3).

2. **Tap the Title field and type the event's name.**

   As you type, the app may suggest automatic completions from previous events.

3. **Tap the Location or Video Call field and type the location or the video call type.**

   The Location screen appears. Tap Current Location to use your current location. Otherwise, type the location or tap an entry in the Video Call list or the Recents list.

4. **If the event will last all day, set the All-Day switch on (green).**

5. **Tap the Starts button to display date and time controls, and then set the start date and time.**

6. **Similarly, tap the Ends button to display date and time controls, and then set the end date and time.**

7. **(Optional) Tap Travel Time and set the travel time, from 5 Minutes to 2 Hours.**

8. **If this event will repeat, tap Repeat, and then tap the interval on the pop-up menu: Every Day, Every Week, Every 2 Weeks, Every Month, Every Year, or Custom.**

   If you tap Custom, the Custom screen appears. Here, you can set up a repetition such as Monthly, On the Second Tuesday. To end the repetition on a specific date, tap End Repeat, tap On Date, and set the date.

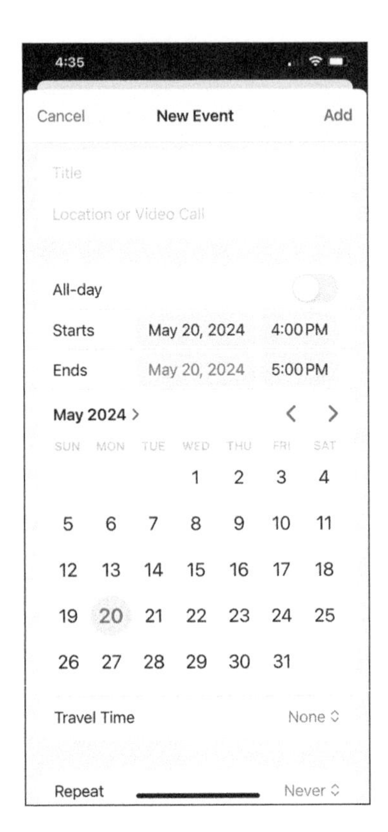

FIGURE 11-3: The screen looks like this just before you add an event to your iPhone.

9. **Tap Calendar and then tap the calendar to which to assign the event.**

10. **To invite people to the event, tap Invitees, and then select the invitees on the Add Invitees screen.**

11. **To set an alert for the event, tap Alert and choose At Time of Event or a time before, such as 5 Minutes Before or 1 Day Before.**

    At that time, your iPhone displays a notification. So does your Apple Watch, if you have one. On the iPhone, you can press the notification and then tap Snooze to get 9 minutes' grace.

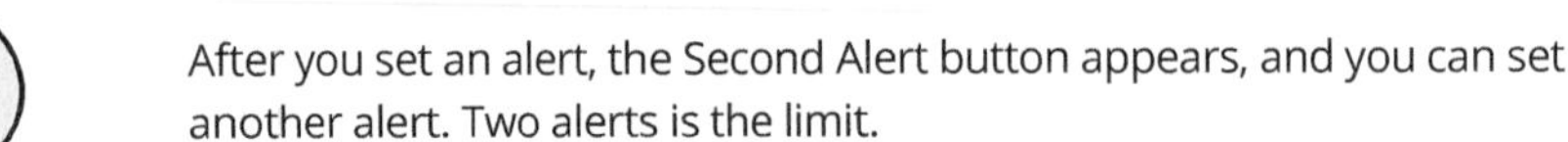

TIP

After you set an alert, the Second Alert button appears, and you can set another alert. Two alerts is the limit.

12. **To control how the event time appears in your calendar, tap Show As, and then tap Busy or Free.**

13. **To attach a file, tap Add Attachment and select the file in the file browser.**

14. **To add a web address, tap URL and type or paste the URL. Double-tap URL and then tap Scan URL to enter a URL via the Live Text feature.**

15. **To add a note, tap Notes and type or dictate the note.**

16. **When you've entered all the event's details, tap Done.**

After creating an event, you can edit it or delete it. Tap the event to display the Event Details screen, and then tap Edit to open it for editing or tap Delete Event to delete it.

Choosing Calendar settings

To make Calendar work your way, tap Settings ➪ Calendar and configure its settings. The following six settings are typically the most useful, but you should also explore the others:

- **Default Calendar:** Tap this button, and then tap the calendar to use for appointments created from other apps, such as from Mail.
- **Time Zone Override:** To make events appear in a specific time zone rather than in your current time zone, tap this button, set the Time Zone Override switch on (green), tap the Time Zone button, and then specify the time zone by typing its name.
- **Sync:** To specify which events to sync, tap this button, and then tap Events 2 Weeks Back, Events 1 Month Back, Events 3 Months Back, Events 6 Months Back, or All Events.
- **Default Alert Times:** Tap this button to display the Default Alert Times screen; tap Birthdays, Events, or All-Day Events; and then tap the timing. For events, you might set 15 Minutes Before; for a birthday, you might set 1 Week Before.
- **Start Week On:** Tap this button, and then tap System Setting or a specific day, such as Monday.
- **Delegate Calendars:** If this button appears, tap it to specify the people you want to allow to manipulate your calendar. Be clear that these people can make changes to your calendar, such as adding and deleting events. If you just want other people to be able to see your events, share your calendar rather than delegating it.

Choosing which calendars to display

By default, Calendars displays the events in each calendar you've added. To specify the selection of calendars, tap the Calendars button at the bottom of the screen, and then tap to place check marks in the circles next to the calendars you want to show or to remove the check marks from the calendars you want to hide. Tap the Hide All button to remove the check marks from all the calendars, or tap the Show All button to place check marks on them all.

The Birthdays calendar shows the birthdays of your contacts in the Contacts app. The Siri Suggestions calendar shows events Siri found in apps, such as Mail and Messages. The Show Declined Events item lets you control whether events you've declined appear in your calendar so you can see what pleasures or nightmares you're missing.

On the Calendars screen, tap the *i*-in-a circle to display the Edit Calendar screen. Here, you can change the name, share the calendar with others by tapping Add Person, change the color by tapping the Color button, enable or disable alerts, and make a calendar public. You can also tap Delete Calendar to delete the calendar. Tap Done when you finish.

Responding to event and calendar invitations

Tap the Inbox button in the lower-right corner of all views except week view to display your Calendar inbox, which collects calendar-related invitations sent to you. Tap the New tab at the top to view your new invitations, where you can tap Accept, Maybe, or Decline to deal with each event invitation; or tap the Replied tab to view invitations to which you've already replied. On either tab, tap an invitation to display its full details.

When someone shares a calendar with you, the Calendar inbox displays a Join invitation. Tap Join Calendar or Decline, as appropriate.

Subscribing to calendars

As well as maintaining your own calendars, you can subscribe to other people's calendars so you can view their events. You can't make changes to subscribed calendars.

To subscribe to a calendar, tap Settings ⇨ Calendar ⇨ Accounts ⇨ Add Account ⇨ Other. On the Add Account screen, tap Add Subscribed Calendar, type or paste the subscription calendar's address, and then tap Next. On the Subscription screen, change the display name if needed, set the Remove Alerts switch on (green), and then tap Save.

Sharing a family calendar

Apple's Family Sharing feature enables you, the family organizer, to share a calendar with up to five other family members. Each member signs in to their iCloud account, and the Family calendar appears in the Calendar app on their Macs, iPhones, and iPads. See Chapter 5 for more information on Family Sharing.

Staying Organized with Reminders

When you need to organize your tasks, turn to the Reminders app. Reminders takes advantage of iCloud to sync your tasks across the iPhone, iPad, Apple Watch, and Mac, enabling you to manage your to-do list from any of your devices. As well as time-based reminders, the app lets you create location-based reminders, so you can set a reminder to trigger when you arrive at (say) the office or when you leave it.

Tap Reminders on the Home screen to launch the Reminders app. Figure 11-4 shows the initial screen: a Search box at the top; a group of categories, such as Today, Scheduled, All, Flagged, and Completed; the My Lists box, which includes the default Reminders list plus any lists you've created; and at the bottom the New Reminder button and the Add List button.

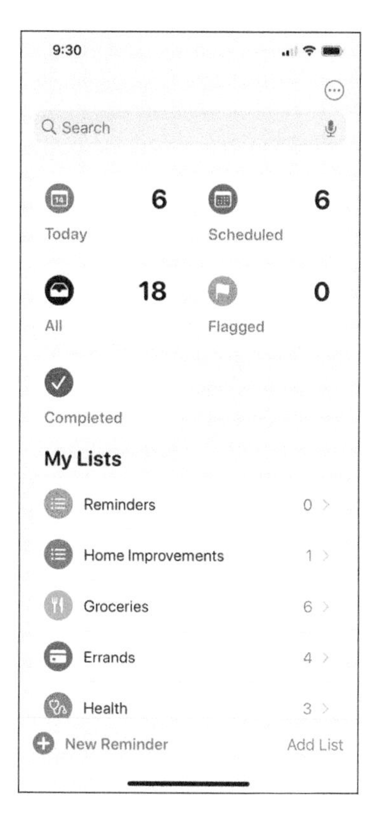

FIGURE 11-4: The initial Reminders screen.

Creating a new reminders list

Reminders starts you off with a single reminders list called simply Reminders. Putting all your reminders into a single list works well only if you have very few reminders. Most likely, you'll want to create other lists.

To create a list, tap Add List in the lower-right corner. On the New List screen, type the list name; tap List Type, and then tap Standard, Groceries, or Smart List. Next, tap the color for the list; tap the icon for the list (tap the upper-left icon to select an emoji instead of an icon); and then tap Done. The new list appears on the first screen, and then opens for editing so that you can add reminders immediately.

Creating reminders

Here's the easiest way to create a new reminder:

1. **On the initial screen (refer to Figure 11-4), go to the My Lists section, and tap the list to which you will add the reminder.**

   The list opens, as shown in Figure 11-5.

2. **Tap the New Reminder button in the lower-left corner.**

   Reminders starts a new reminder on the next line after the last reminder.

3. **Type the text of the reminder, and then tap the *i*-in-a-circle icon to its right to display the Details screen.**

4. **(Optional) Tap Notes and add a note, or tap URL and add a URL. Or add both if you like.**

5. **To set a date for the reminder, set the Date switch on (green), and then tap the date.**

   You can then set a time by setting the Time switch on (green) and using the spin wheels to specify the hour, minutes, and AM or PM.

   You can also tap Early Reminder and then set the early reminder's lead time, such as 15 Minutes Before.

   To create a repeating reminder, tap the Repeat button, and then specify the schedule.

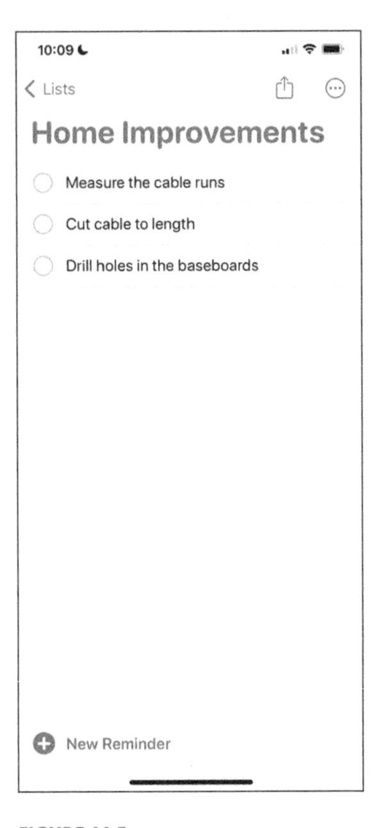

FIGURE 11-5: Tap the New Reminder button to start creating a new reminder in the open list.

6. **To add a tag to the reminder, tap Tags, type the tag on the Tags screen, and then tap Done.**

   Reminders adds the tag, prefacing it with a #, such as #industry. Once you've created a tag, Reminders suggests reusing it on other new items that seem appropriate.

7. **To set a location for the reminder, set the Location switch on (green), and make further selections as explained next.**

   Four location buttons appear: Current, Getting In, Getting Out, and Custom. To use the current location, tap Current. To set the reminder to trigger when you get into your car, tap Getting In, then tap the *i*-in-a-circle on the Getting in the Car button, and specify the location on the Location screen. The Getting Out button works in a similar way to the Getting In button. To specify a different location without involving a car, tap Custom, and then specify the location on the Location screen.

TIP

The Getting In and Getting Out events use Bluetooth. The Getting In event occurs when your iPhone connects to a car with which it is paired. The Getting Out event occurs when your iPhone disconnects from a paired car.

8. **To make the reminder trigger when you're messaging a particular person, set the When Messaging switch on (green), tap Choose Person, and select the person.**
9. **To flag the reminder as important, set the Flag switch on (green).**
10. **To set the reminder's priority, tap Priority, and then tap Low, Medium, or High.**
11. **To move the reminder to a different list, tap List, and then tap the list.**
12. **To add subtasks, tap Subtasks and work on the Subtasks screen.**
13. **To add an image, tap Add Image; tap Take Photo, Scan Document, or Photo Library; and then follow the prompts.**
14. **Tap Done to add the reminder to the list.**

TIP

You can also create a reminder by tapping New Reminder in the lower-left corner of the initial screen. This method uses a slightly different interface but offers the same capabilities. Alternatively, tell Siri to create a reminder for you by saying something like "Hey Siri, remind me to put my iPad in the car at seven o'clock tomorrow morning."

Working with your reminders

To view your reminders, display the appropriate category or list. For example, tap the Today category to view all reminders scheduled for today, or tap the Reminders list to see all the reminders it contains.

To edit a reminder, tap it, tap the *i*-in-a-circle button, work on the Details screen, and then tap Done.

To mark a reminder as done, tap its circle. The circle appears filled in for a moment, and then the reminder disappears.

TIP

Reminders hides your completed reminders by default to reduce clutter. To see all your completed reminders, tap the Completed category on the initial Reminders screen. To see the completed reminders for a list, display the list, tap the ellipsis-in-a-circle button in the upper-right corner, and then tap Show Completed. (To hide the completed reminders, tap the ellipsis-in-a-circle again, and then tap Hide Completed.)

To delete a reminder, swipe left on it, and then tap the Delete button that appears.

To delete an entire list, swipe left on the list in the My Lists list, and then tap the trashcan icon.

Making the Most of Clock

The iPhone's Clock app is straightforward but can be extremely useful, providing world clock, alarm, stopwatch, and timer features. The timer even enables you to stop media playing after a specified period.

World clock

Tap the Clock icon on the Home screen to launch the Clock app, and then tap the World Clock tab at the bottom of the screen. The World Clock screen appears, showing the list of cities you've added (see Figure 11-6).

FIGURE 11-6: Clocking in around the world.

To add another city, tap + (add icon) in the upper-right corner of the screen. On the Choose a City screen, either browse to the city you want or tap in the Search box and start typing its name. Tap the city to add it to the end of your list.

To remove a city, swipe its button all the way left off the list. To reorder the list of cities, tap Edit, drag the city buttons into your preferred order using the handles on their right sides, and then tap Done.

Alarm clock

To set an alarm (or as many alarms as you want), tap the Alarms tab at the bottom of the Clock screen, and then use the controls on the Add Alarm screen to specify the time and choose options:

» **Repeat:** Tap this button to choose which days the alarm should repeat.

» **Label:** Tap this button and type a name to identify the alarm, such as Weekday Waker.

» **Sound:** Tap this button to set the sound and vibration pattern for the alarm.

» **Snooze:** Set this switch on (green) to include a Snooze button that gives you 9 minutes' peace before resuming hostilities.

When you've made your choices, tap Save to save the alarm. On the Alarms screen, enable or disable the alarms, as needed, by setting their switches on (green) or off (white). When one or more alarms are set, a clock icon appears in the status bar of Touch ID iPhone models and in Control Center for all iPhone models.

TIP

Use Siri to set an alarm quickly — for example, "Hey Siri, set an alarm called Workout Purgatory for 4:30AM." Or to turn off an alarm — for example, "Hey Siri, turn off my Workout Purgatory alarm." Or to display the alarms in your future, such as "Hey Siri, show me my alarms."

WARNING

Alarms sound even when your iPhone is set to silent mode and when you have a focus enabled. You can control the alarm volume by choosing Settings ⇨ Sounds & Haptics and setting the Ringtone and Alert Volume slider.

Stopwatch

To use the Stopwatch feature, tap Stopwatch at the bottom of the Clock screen. Swipe left at the top of the screen to switch from the digital stopwatch to the analog stopwatch; swipe right to switch back.

Tap Start to begin timing; tap Lap to record the times of individual laps, if needed; and then tap Stop to finish. Tap Reset when you're ready to reset the stopwatch.

Timers

When you need a countdown timer, tap Timers at the bottom of the Clock screen. Spin the hour, minute, and second dials to set the time. Tap Label and type a descriptive label, such as Bread Timer, to distinguish this timer from others. Tap When Timer Ends, select the sound to play, and then tap Set. You can then tap Start to start the timer running. It appears on the Timers screen showing the countdown and a pause button. You can then tap + (add) in the upper-right corner of the screen to start creating another timer.

When the timer sounds, tap the notification to silence it.

TIP

You can set a sleep timer in the iPhone to shut down music or a video after you've shut down. Set the amount of time you want the iPhone to play, tap When Timer Ends, and then tap Stop Playing. The iPhone will be silenced when the time runs out.

Siri can also set a timer on your behalf. Or you can manually set a timer from Control Center.

IN THIS CHAPTER

» Finding your way with Maps and Compass

» Tracking your stocks, your files, and your stuff

» Working with Calculator and Voice Memos

» Using your iPhone with your car

» Translating text or speech

Chapter 12 Using Maps and Other Helpful Tools

In this chapter, you start using a range of helpful tools that come with your iPhone. The Maps app and the Compass app help you find yourself and your way, while the Stocks app and the Weather app keep you up to date with the markets and meteorological conditions, respectively. The Files app enables you to keep track of your documents, whereas AirTags let you track physical objects. The Calculator app enables you to perform either simple calculations or mathematical and scientific calculations; the Voice Memos app gives you a handy audio recorder; and the Wallet app lets you make payments easily and securely both online and offline.

By using AirDrop and NameDrop, you can share data such as photos or contact cards wirelessly and almost effortlessly. The Home app put your iPhone in control of your smart home appliances, while the CarPlay feature lets you use your iPhone in and with your car. The Measure app provides easy measurement and levelling, and the Translate app helps you to navigate and communicate in other languages. Finally, the Health app enables you to track many aspects of your health information.

Finding Yourself and Your Destination

The Maps app lets you quickly and easily do the following:

- Discover exactly where you are.
- Find nearby restaurants and businesses.
- Get turn-by-turn directions for driving, cycling, or walking from one address to another.
- See real-time traffic information for most urban locations.

Although the Apple Maps app has been upgraded regularly since it replaced Google Maps on Apple devices many years ago and now offers information on public transportation, electric vehicle routing, bicycling directions, indoor maps of malls and airports, and such, some of these features are still not available outside the largest cities. For lists of cities that offer these and other features, visit `www.apple.com/ios/feature-availability`.

TIP

The free Google Maps app is available in the iOS App Store and offers some of the features just mentioned in more cities than Apple Maps. For any high-stakes journey, checking your route on both Apple Maps and Google Maps can be a wise move. The Waze app also is well worth a look.

Finding your current location with Maps

To determine your current location, tap the tracking icon, the arrowhead near the upper-right corner of the Maps screen (and shown in the margin).

You can also can long-press the Maps icon and choose a quick action to open Maps. Your choices are Search Nearby, Send My Location, and Mark My Location.

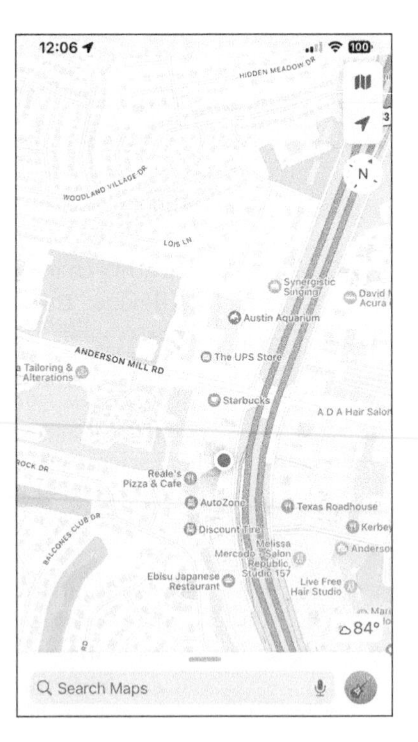

FIGURE 12-1: The blue marker shows your location.

When Location Services is actively finding your location for Maps or any other app or feature, a Location Services indicator (a black arrowhead) appears in the status bar, as shown in Figure 12-1.

HOW DOES MAPS DETERMINE YOUR LOCATION?

The Maps app determines your approximate location by using your iPhone's Location Services, which uses available information from your carrier's cellular network, local Wi-Fi networks (if Wi-Fi is turned on), Bluetooth, and GPS. If you're not using Location Services, turning it off (tap Settings ⇨ Privacy & Security ⇨ Location Services) will conserve your battery. If Location Services is turned off when you tap the Tracking icon, iOS prompts you to turn it back on. Note that Location Services may not be available in all areas at all times.

REMEMBER

If you tap or drag the map, your GPS continues to update your location but won't recenter the marker, which means that the current location indicator can move off the screen. Tap the tracking icon (the arrowhead near the upper-right corner) to move the map so that the current location marker appears in its center again.

When you tap the tracking icon, it turns solid gray (refer to Figure 12-1), which indicates that your current location is in the middle of the map. If you tap, drag, rotate, or zoom the map, moving the current location indicator from the center of the screen, the tracking icon turns white with a dark gray outline. So, if the icon is all gray, your current location is currently in the middle of the screen; if the icon is white with a gray outline, your current location is elsewhere.

REMEMBER

In Satellite view, the tracking icon appears in light gray against a dark background, When your current location is in the middle of the screen, the icon fills with light gray.

Finding a person, place, or thing

To find a person, place, or thing with Maps, tap the search field at the bottom of the screen, and then type what you're looking for. You can search for addresses, zip codes, intersections, towns, landmarks, and businesses by category and by name, or combinations, such as *New York, NY 10022; pizza 60611;* or *Auditorium Shores Austin TX.*

TIP

If the letters you type match contacts who have street addresses, those contacts appear in a list below the search field. Tap a name to see a map of that contact's location.

When you finish typing, tap Search. After a few seconds, a map appears. If you searched for a single location, it's marked with a single bubble. If you searched for a category (*Cambodian cuisine Oakland CA*, for example), you see multiple bubbles, one for each matching location (Cambodian restaurants in and around Oakland), as shown in Figure 12-2.

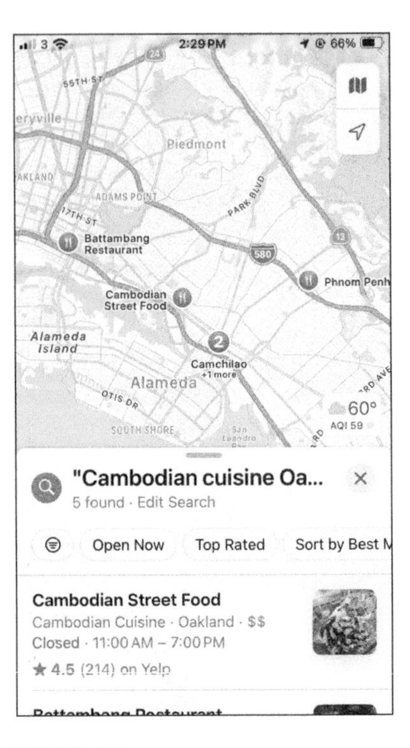

FIGURE 12-2: Search for *Cambodian cuisine Oakland CA* and you see bubbles for all Cambodian restaurants in Oakland.

The oval buttons near the bottom of Figure 12-2 — Open Now, Top Rated, and so on — are filters. Tap one to see only restaurants that match the filter. Or tap the three lines icon to the left of Open Now to see additional filtering options, including price and amenities.

If a bubble on the map has a number rather than a name, you'll see that number of places when you zoom in.

Whenever you start a search in Maps, a bunch of "find nearby" icons pop up below the search field, including Restaurants, Fast Food, Food Delivery, Groceries, Gas Stations, and Shopping Centers. Tap an icon to see nearby establishments in that category.

Views, zooms, and pans

Maps offers four view: explore view, driving view, transit view, and satellite view. Figure 12-2 shows explore view, while the left screen in Figure 12-3 shows the satellite view of the same area. Select a view by tapping the view icon above the current location icon to display the Choose Map dialog (see Figure 12-3, right), tapping the map view you want, and then tapping X (close) in the upper-right corner of the dialog.

Transit view is just explore view with added information about bus and train routes and stations. At present, it's available in larger cities, with additional cities added regularly. Driving view looks like explore view but includes traffic information.

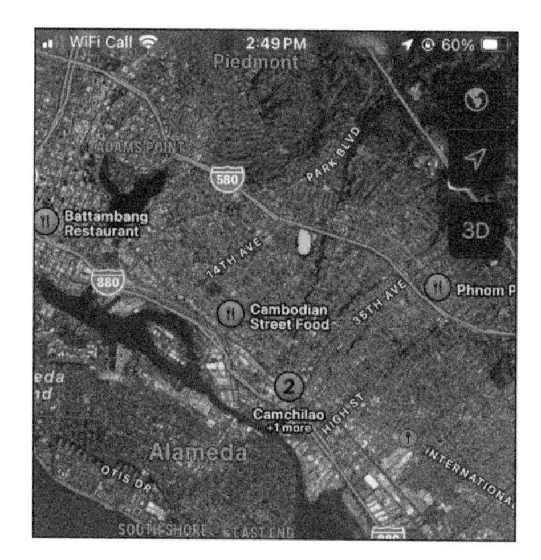

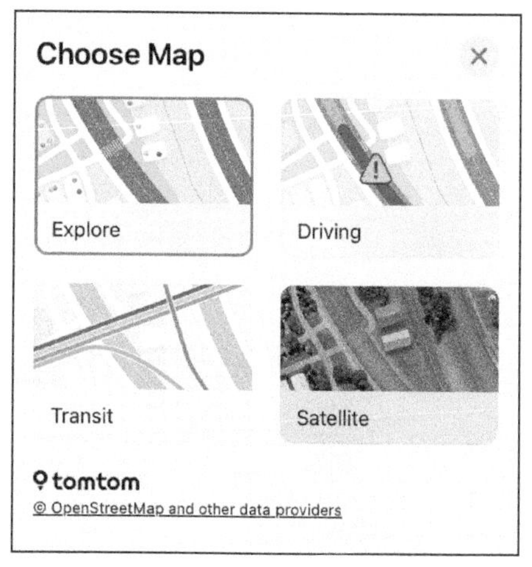

FIGURE 12-3: Satellite view of the map shown in Figure 12-2 (left) with the Choose Map dialog (right).

TIP

Satellite view can display labels (for place names) and traffic information if you want; tap the Satellite thumbnail in the Choose Map dialog, tap the ellipsis-in-a-circle button that appears, and then tap the Show Labels item or the Show Traffic item, placing or removing a check mark on the item, as appropriate.

In all views, you can zoom in or out to see either more or less of the map, or scroll (pan) to see what's above, below, or to the left or right of what's on the screen:

- **Zoom out:** Pinch the map or double-tap using two fingers. To zoom out even more, pinch or double-tap using two fingers again.
- **Zoom in:** Unpinch the map or double-tap (the usual way — with just one finger) the spot you want to zoom in on. Unpinch or double-tap with one finger again to zoom in even more.

  An *unpinch,* sometimes called a *spread,* is the opposite of a pinch. Start with your thumb and a finger together and then spread them apart.
- **Rotate:** Place two fingers on the screen and then rotate them. The compass in the upper-right corner of the screen rotates in real time as you move your fingers.
- **Scroll:** Flick or drag up, down, left, or right with one finger.

Timesaving map tools: Contacts, Recents, Guides, and Favorites

The Maps app offers four tools that can save you from having to type the same locations over and over: Contacts, Recents, Guides, and Favorites.

Contacts

Maps and contacts go together like peanut butter and jelly. To see a map of the street address for a contact you've added, just type the first few letters of the contact's name in the search field at the bottom of the screen. Now tap the contact's name in the list, and a bubble drops over the person's house (on the map, of course) and the Contacts button and the Directions button appear. Tap Contacts to display the person's contact card in an overlay; tap Directions and to display step-by-step directions.

REMEMBER

If you have multiple addresses for a contact, you'll see multiple entries — one for each address — in the overlay. Tap the address you want.

After you find a location by typing an address in Maps, you can add that location to one of your contacts or create a contact with a location you've found. To do either, tap the location's bubble on the map, and then tap its info bubble to display its Info screen. Tap the share icon near the top, and then tap Create New Contact or Add to Existing Contact.

REMEMBER

You can work with contacts in three ways. The first way is to tap the Contacts icon on the Home screen. The second way is to tap the Phone icon on the Home screen and then tap the Contacts icon at the bottom of the screen. Finally, you can use the Info screen for a location in the Maps app, as just described.

When you're finished, tap the little x-in-a-circle in the top-right corner of the overlay to return to the map.

Recents

The Maps app automatically remembers every location you've searched for in its Recents list (unless you've cleared it, as described next). To see this list, swipe upward on the gray bar above the search field, and then scroll down until you see Recents.

To see more than the last three searches, tap More. The Recents overlay appears, showing several time-based categories such as Today, This Week, This Month, and Older. Each category has its own Clear button. To clear an item from the Recents list, either swipe the item partway from right to left and then tap its Delete button, or swipe all the way from right to left, making the Delete button grow and engulf the item. To clear an entire category, tap Clear to the right of the category heading.

When you're finished, swipe downward on the gray bar above the search field to minimize the overlay and return to the map.

Guides

Guides are collections of places you can create and share with others. There are also guides created by "brands you trust" (at least according to Apple). To see the curated guides (currently available only in select cities), tap the search field, scroll down, and then tap Explore Guides.

Guides update automatically, so if you create one and share it with others, their guide will update when you add or delete locations from it.

To create a guide, swipe upward on the gray bar above the search field, and then tap New Guide. Type a name for the guide, select an opening photo by tapping the camera icon, and then tap Create. Tap your new guide to do the following:

- **Add a new place.** Open an overlay where you can search for a place or choose one from your Recents list by clicking its + button.
- **Edit the guide.** Change the name of the guide, add or change its photo, or delete it.
- **Share.** Display a share sheet that provides the standard share options — Mail, Messages, AirDrop, Notes, and Reminders — plus other available means of sharing, such as Facebook or X (formerly Twitter).
- **Change the sort order of the list.** Choose Name, Distance, or Date Added.

When you're finished with your guide, tap the little x-in-a-circle to close it.

Favorites

Favorites in the Maps app, like bookmarks in Safari, let you return to a location without typing a single character. (Earlier versions of the Maps app had bookmarks instead of favorites.)

If you've already searched for an address, you can make it a favorite by scrolling down, or by tapping the ellipsis (. . .), and then tapping Add to Favorites.

You can also save a location as a favorite by dropping a *pin* (a temporary favorite) on the map to mark the location. To drop a pin, press with one finger until a bubble with a pin appears. To reposition a pin, press the bubble and drag it on the map. When you lift your finger, you'll see an info bubble with the location of the pin on an overlay near the bottom of the screen (if Maps can figure it out). Scroll down and tap Add to Favorites to save the location as a favorite.

After you add a location to Favorites, you can recall it at any time by first swiping up on the gray bar above the search field, scrolling down to Favorites, and then tapping the icon for the favorite.

TIP

Create favorites for your home and work addresses so that you can use these addresses easily. Also create zip code favorites for your home, work, and other locations you frequently visit. Then when you want to find businesses near any of those locations, you can choose the zip code favorite and add what you're looking for, such as *78729 pizza*, *60611 gas station*, or *90201 Starbucks*.

To share a favorite, tap More, swipe left on the favorite, and then tap Share. Similarly, to delete a favorite, tap More, swipe left, and then tap Delete.

When you're finished using Favorites, swipe down on the overlay to return to the map.

Smart map tricks

The Maps app includes route maps and driving directions; real-time traffic info; directions for walking, bicycling, and public transit; and even 3D views.

Get route maps and driving directions

You can get route maps and driving directions to any location from any other location in a couple of ways.

Here's the first method: If a marker is already on the screen, tap the bubble with the name of the location, and then tap the Directions button in the overlay to get directions to or from that location.

The second method applies to search results and items from lists:

1. **In your Favorites list or Recents list, search for a destination or tap one.**
2. **Tap the blue Directions button.**

   An overlay slides up from the bottom of the screen with the details of your trip shown as a list of stops, the estimated trip duration, and a big green Go button for each route (see Figure 12-4, left). But, before you tap Go . . .
3. **If you need directions from someplace other than your current location or you want to change your destination, tap My Location or the destination stop, and then type or select a new location on the Change Stop screen.**
4. **If you want to add a stop, tap Add Stop, and then specify the stop on the Add Stop screen.**

   Figure 12-4, right, shows a journey with two stops between the start point and the destination.

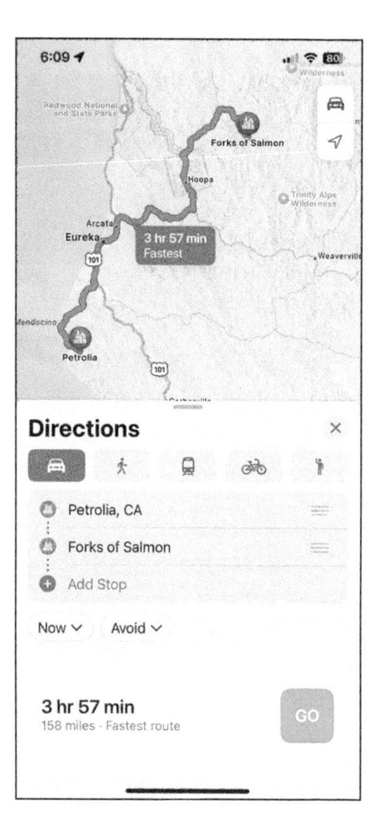

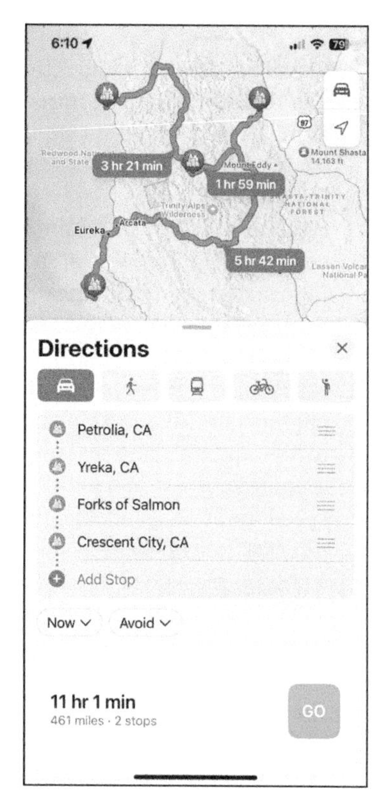

FIGURE 12-4: Maps shows the route to your destination (left). Tap the Add Stop button to add one or more stops along the way (right).

5. **To change the order of your stops, tap the handle on the right side of a stop and drag up or down.**

   For example, if you want to reverse the journey, drag the first stop to last place and the last stop to first place.

TIP

   To remove a stop, swipe left on its button, and then tap the Delete button that appears.

6. **If Maps suggests several possible routes, select one by tapping its line or the time bubble attached to the line.**
7. **Tap the Go button near the bottom of the screen to receive turn-by-turn driving directions, as shown in Figure 12-5.**
8. **To see the next step in the directions, swipe the instructions near the top of the screen from right to left; to see the preceding step, swipe from left to right.**
9. **When you're finished with the step-by-step directions, tap the big red End button near the bottom of the screen to return to the regular map screen.**

Visual step-by-step directions work well, but you'll also hear audible turn-by-turn directions similar to what you'd find on a dedicated GPS device. You know, where some friendly male or female voice states instructions such as "turn right in 900 feet on Main Street." And while you're using step-by-step directions, traffic information is automatically displayed on the map, as described in the next section.

Maps also gives you lane guidance where appropriate, displaying arrows to guide you to the optimal lane or lanes as you approach a multilane interchange. When the current speed limit is available, Maps displays this information on-screen as well. If you don't want to see speed limits on-screen, choose Settings ➪ Maps, go to the Directions box, tap Driving to display the Driving screen, and then set the Speed Limit switch off (white). While you're on the Driving screen, set the Compass switch on (green) if you want to display the compass to help you navigate.

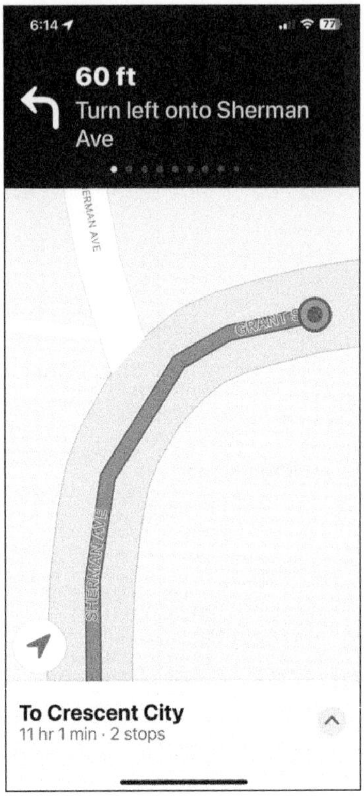

FIGURE 12-5: Step-by-step driving directions point you in the right direction.

Get traffic info in real time

When you're not using step-by-step directions, you can find out the traffic conditions for whatever map you're viewing by tapping the current view button in the upper-right corner and then tapping Driving in the Choose Map dialog. When you do this, major roadways are color-coded to inform you of the current traffic speed. Maps also displays markers for roadwork, wrecks, and other disruptions. Tap a marker to learn the details of the bad news.

Here's the key to those colors:

- **Green:** More than 50 miles per hour
- **Orange:** 25 to 50 miles per hour
- **Red:** Under 25 miles per hour
- **Gray:** No data available at this time

WARNING

Traffic info isn't available in every location, but the only way to find out is to give it a try. You may have to zoom in to see the color codes; if they don't appear, assume that traffic information doesn't work for that particular location.

Get directions for walking, bicycling, public transportation, and ride share

For step-by-step directions, tap the appropriate icon under the Directions heading: drive, walk, transit, cycle, or ride share (from left to right; refer to Figure 12-4).

Get 3D views

The Maps app also offers three-dimensional views for most metropolitan areas. When available, you'll see a 3D button below the current location icon. When 3D is enabled, you navigate and zoom as described earlier in the chapter. To change the camera angle, drag up or down on the screen with two fingers.

Do more on the Info screen

If a location has a little bubble near its name (refer to the Cambodian chow in Figure 12-2), you can tap the bubble to see its Info screen. Here, you can get directions to or from the location, add it to your favorites or to an existing contact, or create a new contact. You can also

- Tap the phone number to call the location.
- Tap the email address to launch the Mail app and send an email to the location.
- Tap the URL to launch Safari and view the location's website.

You can long-press almost any identified location on the map — be it a landmark, a restaurant, a store, a park, or whatever — and choose Directions, Call, Open Homepage, or Share Location from the Quick Actions list.

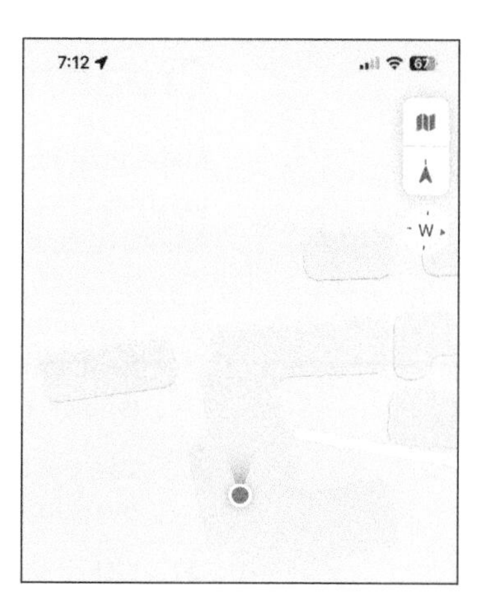

FIGURE 12-6: The cone indicates the direction your iPhone is facing — in this case, west.

Find your direction

Tap the tracking icon (near the upper-right corner) two times, and the location marker displays a cone that indicates the direction your iPhone is facing, as shown in Figure 12-6.

When the map is in compass mode, the current location icon in the upper-right corner grows a little tip (as shown in the margin), to let you know that you're using compass mode. If you rotate to face a different direction while Maps is in compass mode, the map rotates in real time to display the direction you're facing.

Dude, where's my car?

Maps automatically drops a pin whenever you park your car so you can find it later. The feature works with any car that connects to your iPhone via Bluetooth or CarPlay. If yours does, whenever you disconnect the phone and leave your car, the location will be marked with a parked car pin.

To enable the Show Parked Location feature, tap Settings ➪ Maps and set the Show Parked Location switch on (green). You can then find your parked car by either tapping the search field and choosing Parked Car from the suggestions list or asking Siri, "Where is my car parked?"

Indoor maps

Maps includes indoor maps of some large public spaces such as airports and shopping malls. Alas, they're still few and far between, but there should be more of them by the time you read this. For a list of currently available indoor maps, visit `www.apple.com/ios/feature-availability`.

Creating and using offline maps

Maps enables you to download sections of maps to your iPhone so that you can use them even when your iPhone has no internet connection. To download a map, follow these steps:

1. **Tap your picture or monogram (your initials) to the right of the Search field, and then tap Offline Maps to display the Offline Maps screen (shown on the left in Figure 12-7).**
2. **Tap the Download New Map button to display the Download New Map overlap.**
3. **Either search to find the location or tap an item in the Recent Searches list.**

   The location appears, with a map frame across the middle part of it (see the right screen in Figure 12-7).

4. **Drag the side handles or corners of the map frame to select the area you want.**

   You can also drag the map about within the frame, unpinch to zoom in, or pinch to zoom out.

5. **Tap the Download button.**

   Maps downloads the map and stores it.

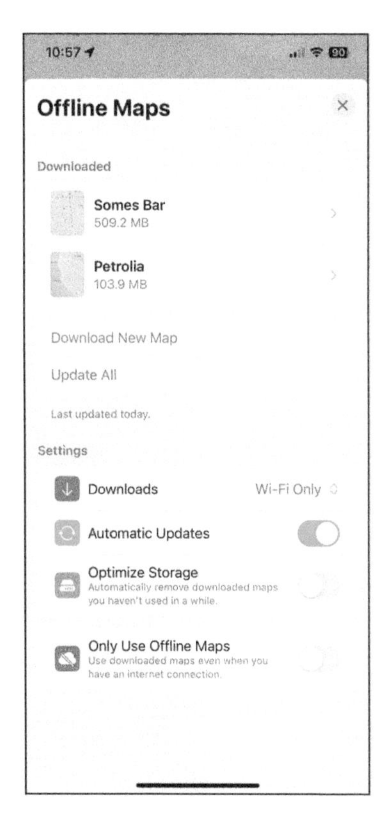

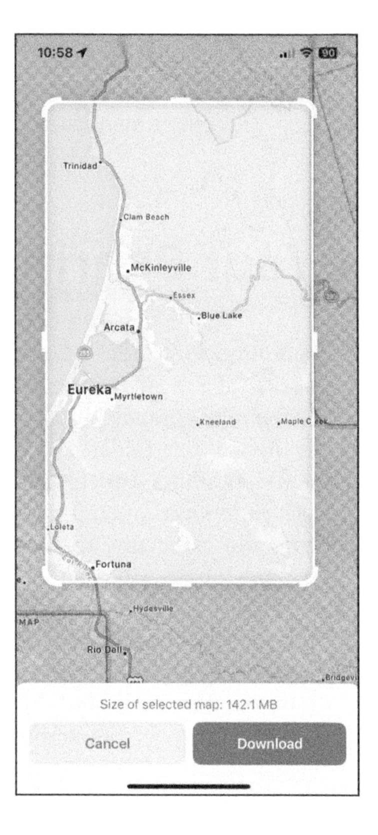

FIGURE 12-7: On the Offline Maps screen (left), tap the Download New Map button and then select the area for the map (right).

To view one of your offline maps, tap your picture or monogram to the right of the search field, and then tap Offline Maps. On the Offline Maps screen, tap the map you want. Its control screen appears, from which you can display the map, rename it (tap the pencil icon), or delete it (tap the Delete Map button).

In the Settings section of the Offline Maps section, you can configure these four settings:

- **Downloads:** Choose Wi-Fi Only unless you have a generous or unlimited cellular data plan — in which case, choose Wi-Fi + Cellular.
- **Automatic Updates:** Set this switch on (green) to have Maps check for and download updates automatically.
- **Optimize Storage:** Set this switch on (green) to have Maps automatically get rid of downloaded maps you haven't used for a while.
- **Only Use Offline Maps:** Set this switch on (green) to make Maps use only your offline maps even when your iPhone has an internet connection.

Getting Your Bearings with Compass

Your iPhone includes a Compass app, which works like a magnetic needle compass. Launch the Compass app by tapping its icon in the Utilities folder on the Home screen. If iOS prompts you to allow Compass to use your location, tap Allow While Using App, and make sure the *Precise: On* readout appears.

You may be instructed to wave your phone around in figures of eight to calibrate the compass, but then you'll see the direction you're facing on the screen, as shown in Figure 12-8.

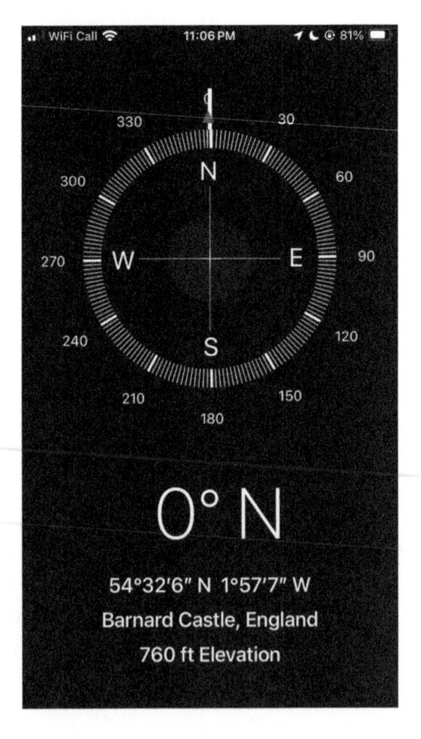

FIGURE 12-8: The Compass app says the iPhone is pointing due north.

TIP

As with a physical compass, you should hold your iPhone level to get an accurate bearing. To make sure the iPhone is level, line up the small crosshair in the floating gray circle with the fixed larger crosshair in the middle of the compass circle.

Herding Your Stocks

The Stocks app enables you to track your choice of stocks and indexes. Each time you open the Stocks app by tapping its icon on the Home screen, it displays the latest prices, with two provisos:

- » The quotes are provided in near real-time.
- » The quotes are updated when your iPhone has an internet connection.

The first time you open Stocks, you see information for a group of default stocks, funds, and indexes. You can't see them all on the screen at once, so flick upward to scroll down. Or swipe up on Top Stories from Apple News at the bottom of the screen to see current financial news items.

The Stocks widget also appears by default in today view. If you don't see Stocks, display Notification Center by swiping right on the first Home screen page, scroll to the bottom of Notification Center screen, and then tap the Edit button. Now tap the add icon (+ in a green circle) next to Stocks to add the widget, and then tap the Done button.

Adding and deleting stocks, funds, and indexes

Your chance of owning the default group of stocks, funds, and indexes shown in Stocks is slim, so you'll want to customize the selection.

Here's how to add a stock, a fund, or an index:

1. **In the search field at the top of the stocks list, type the name of the stock, fund, or index.**

   Swipe down if you don't see the search field.

2. **Tap the company, index, or fund you want to add.**

   The details for the item appear, so you can check you've picked the right one.

3. **Tap the ellipsis (. . .) button, and then tap Add to Watchlist.**

   To return to the main screen without adding the stock, fund, or index to your watchlist, tap the little *x*-in-a-circle near the upper-right corner.

4. **Repeat Steps 1 through 3 until you've finished adding stocks, funds, and indexes.**
5. **Tap Done to return to the main screen.**

Here's how to re-order your list and remove items:

1. **Tap the ellipsis button in the upper-right corner of the main Stocks screen, and then tap Edit Watchlist.**

   Stocks opens the watchlist for editing.

2. **To change the order, drag items up and down by the handles on their right side.**
3. **To remove an item, tap the - icon to the left of its name, and then tap the Remove button that appears to the right of the name.**
4. **When your list is the way you want it, tap the Done button.**

Viewing the details for a stock, fund, or index

To see the details for a stock, a fund, or an index, tap its name. Swipe up to see additional information; tap the little x-in-a-circle near the upper-right corner to return to the main screen.

To look up additional information about a stock at Yahoo.com, first tap the stock's name to display the details, and then tap the Yahoo! Finance icon in the lower-left corner of the screen. Safari launches and displays the Yahoo.com finance page for that stock.

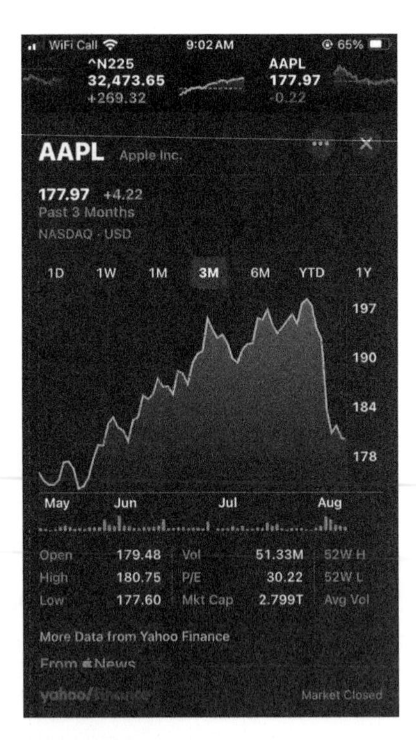

FIGURE 12-9: The three-month performance chart and related news for Apple.

Charting a course

When you tap a stock, a fund, or an index, its performance chart appears. At the top of the chart, you see a bunch of numbers and letters: 1D, 1W, 1M, 3M, 6M, YTD, 1Y, 2Y, 5Y. These stand for 1 day; 1 week; 1, 3, and 6 months; year to date; and 1, 2, and 5 years, respectively. You might also see other periods, such as 10Y and ALL. Tap one of them to make the chart show that period of time. (In Figure 12-9, 3M is selected.)

From here, you can do the following:

- Touch any point in time to see the value for that day.
- Use two fingers to touch any two points in time to see the difference in values between those two days.
- Swipe up to see news about the stock, fund, or index.
- Tap the little *x*-in-a-circle near the upper-right corner to return to the main screen.

By default, the Stocks app displays the change in an item's price in dollars. You can instead see the change expressed as a percentage or as the market capitalization. Tap the number in the colored box next to any stock (green boxes are positive; red boxes are negative) to cycle the display for all items through dollar change, percent change, and market cap. For example, if your stocks, funds, and indexes are currently displayed as dollars, tapping any one of them switches them all to percent — and tapping again switches them all to market cap.

Weather Watching

The Weather app provides you with the current weather forecast for the city or cities of your choice. By default, you see a daily forecast for your current location at the bottom of the screen, as shown in Figure 12-10, with the hourly forecast above.

TIP

Swipe the hourly forecast left or right to display later or earlier hours. To see more detail about a day in the 10-Day Forecast list, tap it. The details screen appears, usually showing a temperature chart for the day. You can then tap the v (downward caret) pop-up menu on the right side and tap Temperature, UV Index, Window, Precipitation, Feels Like, Humidity, Visibility, or Pressure to see a chart for that metric. Tap X in the upper-right corner to return to the main screen.

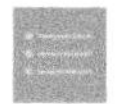

To add a city, first tap the info icon in the bottom-right corner (and shown in the margin) to display the Info screen. Next, tap in the search field at the top of the screen; type a city, a zip code, or an airport location; and then tap the Search button in the bottom-right corner of the screen. Finally, tap the name of the found city to display its details, and tap Add in the upper-right corner if you want Weather to remember this city. Add as many cities as you want this way.

To delete a city, tap the info icon. Swipe left on the city's name, and then tap the delete icon (trashcan) that appears to the right of its name. Alternatively, swipe left on the city's name, but keep swiping until the delete icon takes over the entire width of the screen.

To change the order of the cities, long-press a city until its button bulges slightly, and then drag it up or down. The other cities move out of the way to make space for it.

Tap the more icon (ellipsis in a circle) in the upper-right corner to choose between Fahrenheit and Celsius, to enable inclement weather notifications, or to change the units displayed for wind, precipitation, pressure, and distance.

When you're finished, tap any city to view its current weather.

When you've added one or more cities to Weather, you can switch between them by flicking your finger across the screen to the left or the right.

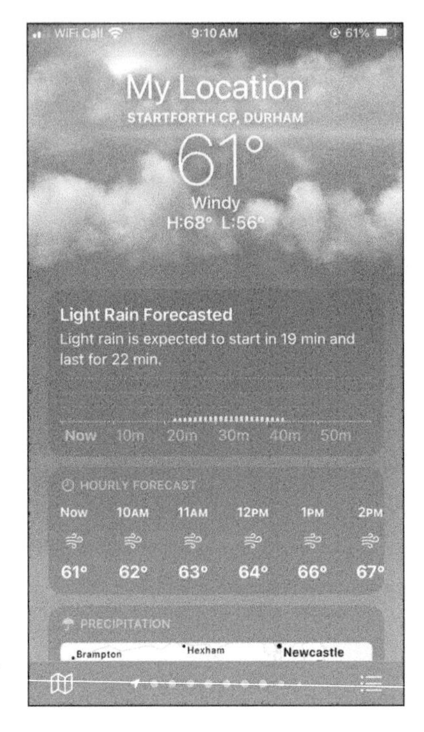

FIGURE 12-10: The hourly and daily local forecasts for Startforth.

TIP

See the Location Services arrow and nine little dots centered at the bottom of the screen in Figure 12-10? The white Location Services arrow means you're currently viewing local weather; the nine dots denote nine other cities in the list.

To see an overlay with temperature, precipitation, air quality, or wind, tap the show map icon (folded map) in the lower-left corner of the screen. Tap the overlay icon (stack of diamonds) to display a pop-up menu that provides the choices Precipitation, Temperature, Air Quality, and Wind; tap the one you want to view. Tap Done in the upper-left corner to return to the previous screen.

Keeping Track of Documents with Files

The Files app lets you see and manage files from third-party cloud storage services such as Dropbox and Google Drive as well as files stored in iCloud.

Files has three tabs at the bottom of the screen:

- **Recents:** This tab displays files you've modified recently.
- **Shared:** This tab displays files shared with you.
- **Browse:** This tab lets you browse files and folders on your iCloud Drive and other cloud-storage services, and view and restore files deleted from your iPhone in the past 30 days.

The Browse tab has three subsections:

- **Locations:** Tap On My iPhone to see files stored locally on your device; tap iCloud Drive to view files stored in iCloud; or tap Recently Deleted to view recently deleted files.

  To add a third-party cloud storage service, first install its app (Dropbox, Google Drive, and so on) from the App Store. Now, tap Files on the Home screen, tap Browse at the bottom of the screen, tap the ellipsis-in-a-circle in the top-right corner of the screen, and then tap Edit. Set the switch on (green) for each service you want to add to Files, and then tap Done.

  Other options for files and folders include the following:

  - Tap a folder to view its contents.
  - Long-press a file or folder for additional options, such as Download, Get Info, Copy, Duplicate, Tag, Share, and Compress.
  - Tap items with a cloud-and-arrow icon to download them to your iPhone.
  - When viewing a PDF or an image file, tap the Markup icon in the upper-right corner (and shown in the margin) to annotate the file using the Markup tools.
- **Favorites:** To add a folder to the Favorites section, long-press the folder icon, and then choose Favorite. The folder must be store on your iPhone or in iCloud Drive. You can't add individual files to Favorites.
- **Tags:** If you use macOS Finder tags, they'll appear in the Tags section. Tap a tag here to see all available files marked with that tag. To tag a file, long-press it, tap Tags, tap the tags you want to apply, and then tap the Done button.

TIP

All three tabs — Recents, Shared, and Browse — have a search field at the top. When you can't find a file or folder by browsing, search instead.

Keeping Track of Stuff with AirTags

When you want to keep track of a physical object in the real world rather than a virtual object in a file system, you can attach an AirTag to that object. An *AirTag* is a Bluetooth device the size of a thick coin that enables you to track it using Apple's Find My app on your iPhone, iPad, or Mac. You can buy AirTags from the online Apple Store (`https://store.apple.com/`), from a bricks-and-mortar Apple Store, or from various other retailers.

To set up an AirTag, bring it and your iPhone close together, and then tap the Connect button in the AirTag dialog that iOS displays. Follow through the setup procedure to the Name AirTag dialog (see Figure 12-11), and then assign a descriptive name. The list offers built-in names, such as Backpack and Keys, but you can also tap Custom Name and type the name you want. When the AirTag Is Linked to Your Apple ID dialog appears, read the restrictions (essentially, no stalking), and then tap the Agree button.

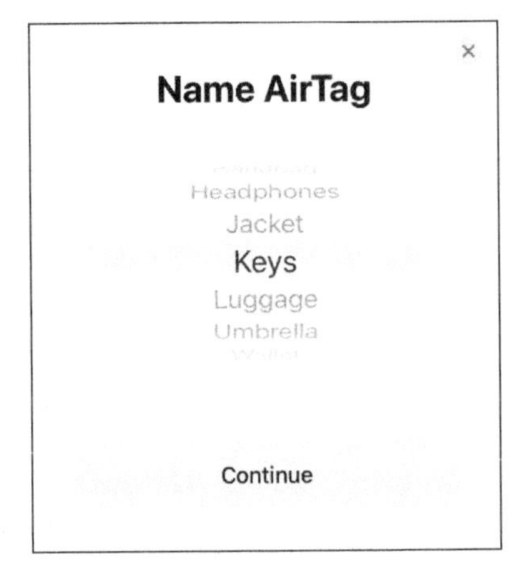

FIGURE 12-11: Assign the AirTag a name describing the object it will track.

When you mislay a tagged item and need to track it down, open the Find My app from the Home screen, and then tap the Items tab at the bottom. The map shows the location of your found item (see the left screen in Figure 12-12). Tap the item you're looking for to display the screen of actions you can take (see the right screen in Figure 12-12). You can then:

- » Tap the Play Sound button to play a sound on the AirTag.
- » Tap the Find button to display a tracking screen showing the distance to the AirTag, enabling you to close in on it, again playing a sound if needed.
- » Tap the Add Person button to start sharing this AirTag with someone else.
- » Set the Notify When Found switch on (green) to have Find My notify you when it locates the AirTag.
- » Tap the Notify When Left Behind button to display the Notify When Left Behind screen, and then set the Notify When Left Behind switch on to have Find My warn you when you seem to have left the AirTag behind. In the Notify Me, Except At area, you can create a list of places where you don't want this notification.

» Tap the Enable button in the Lost Mode area to turn on Lost Mode for the AirTag. Lost Mode locks the AirTag to your Apple ID so nobody else can use it, makes the AirTag display a message when someone else finds it (encouraging them to contact you), and notifies you when it has been found.

» Tap the Rename Item button to rename the AirTag.

» Tag the Remove Item button to unlink the AirTag from your Apple ID so that someone else can use it.

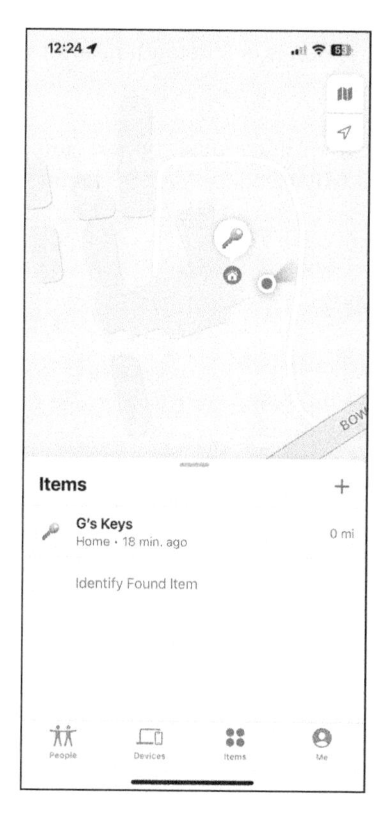

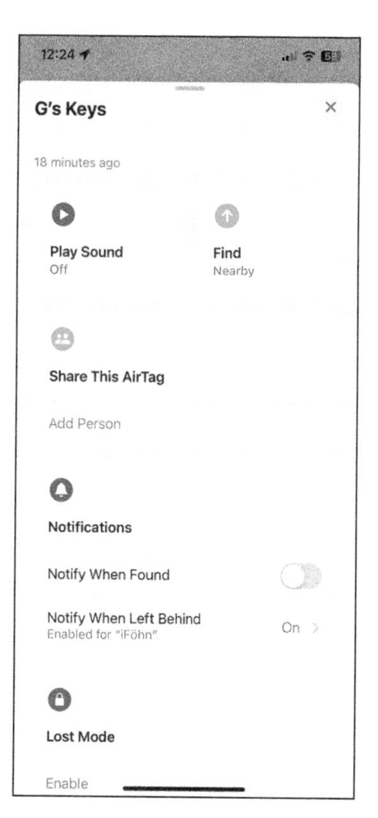

FIGURE 12-12: Use the Find My app to locate a missing AirTag (left). You can then play a sound on the AirTag or track it down (right).

Using Calculator

When you need to perform calculations, open the Utilities folder on the Home screen and tap the Calculator icon. Calculator opens in its basic format, providing buttons for addition, subtraction, multiplication, division, percentages, and changing a number's sign (+ to – or vice versa).

When you need more calculating power, rotate your iPhone to landscape orientation. Calculator switches to a scientific calculator that can calculate everything from cube roots to logarithms. The scientific calculator also has memory functions for storing values.

Working with Voice Memos

The Voice Memos app lets you use your iPhone as a voice recorder. You can record using the iPhone's built-in microphones, a Bluetooth headset mic, or an external microphone.

To launch Voice Memos, open the Utilities folder on the Home screen, and then tap the Voice Memos icon. If you will use Voice Memos frequently, consider moving the Voice Memos icon to a Home screen page for quicker access.

Making a recording

Tap the red record icon in the lower part of the screen to start recording. The waveform that moves across the screen as Voice Memo detects sounds helps you gauge the recording level. A timer on the screen indicates the length of your recording session. Tap the red button again to stop and save the recording.

TIP

For best results, speak in a normal voice. To adjust the recording level, move the microphone closer to or farther from your mouth.

Listening to recordings

After you've finished making a recording, a list of all your recordings pops up in chronological order, with the most recent on top, as shown in Figure 12-13. Tap the recording you want to play back and then tap the play button.

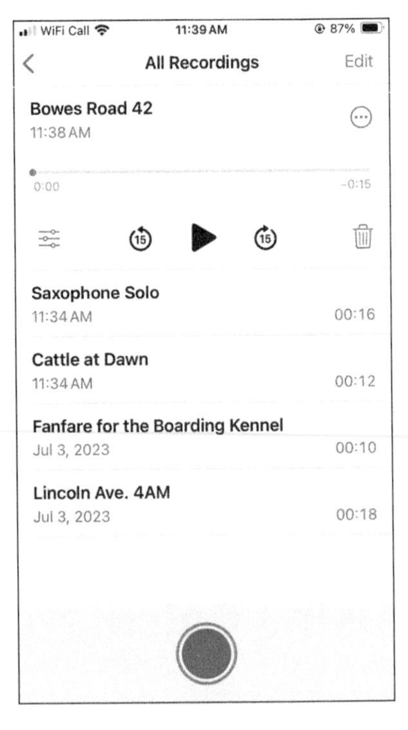

FIGURE 12-13: Tap the voice memo you want to play.

TIP

You can drag the playhead along the scrubber bar to move to any point in the memo. Alternatively, tap the circular 15 buttons to move 15 seconds back or forward.

To choose playback options, tap the options icon (shown in the margin), and then use the three controls in the Options dialog:

- » **Playback Speed:** Drag the slider along the tortoise-hare continuum to slow down or speed up playback.
- » **Skip Silence:** Set this switch on (green) to have Voice Memos skip lightly over silences during playback.
- » **Enhance Recording:** Set this switch on (green) to have Voice Memos attempt to make the recording sound better by reducing background noise, balancing the audio levels, and boost the clarity of voices.

Trimming a recording

To trim unwanted parts off the start or end of a recording, tap the recording, tap the ellipsis-in-a-circle icon, and then tap Edit Recording. On the Edit screen, tap the trim icon (shown in the margin), and then drag the yellow start marker and end market to specify the audio segment you want to keep. Tap the play icon to play back the segment to make sure it's right; when it is, tap the trim icon, and then tap the Save button.

TIP

If you've trimmed a recording and have second thoughts, shake your iPhone, and then tap Undo. But after you've tapped Done, your edits are permanent. To keep the original recording unchanged, tap the three-dot icon and choose Duplicate to create a copy of the recording, and work on the copy instead of the original.

Adding to a recording

You can add to a recording either by replacing part of it or by appending a new section. To get started, tap the recording, tap the ellipsis-in-a-circle icon, and then tap Edit Recording. The Edit screen appears. To replace content, drag the playhead to the beginning of what you want to replace, tap the Replace button, and speak your part. To append content, drag the playhead to the end of the recording, and then tap the Resume button (which appears in place of the Replace button when the playhead is at the end of the recording).

Renaming a recording

When you save a recording, Voice Memos gives it either the uninspiring name New Recording or the location where you made it. (You can control location settings in the Settings app.) To rename a recording, tap it in the list of recordings, and then type the new name.

Sharing and syncing recordings

To share a recording, tap the ellipsis-in-a-circle icon, and then tap Share. You can then share the recording via Mail, Messages, AirDrop (discussed later in this chapter), or another means.

You can sync voice memos to your Mac through Finder on recent versions of macOS, or to your PC or a Mac with an older version of macOS by using iTunes, as described in Chapter 3. Recordings you sync from your iPhone to your Mac or PC appear in the Voice Memos playlist.

From the menu opened by tapping the ellipsis-in-a-circle icon, you can tap Favorite to mark the voice memo as a favorite or tap Save to Files to save the voice memo to a destination in the Files app.

You can also turn a recording into a ringtone. Email the recording to yourself, download it to your Mac or PC, change the file extension to .m4r, and then put the recording in the Music app or iTunes, respectively.

Deleting a recording

To delete a recording, swipe partway left on the recording's button in the list of recordings, and then tap the delete icon (trashcan). Alternatively, swipe the recording's button all the way to the left until the delete icon (trashcan) consumes it.

Choosing settings for recordings

Choose Settings ➪ Voice Memos. You can then configure these settings:

- **Clear Deleted:** To specify when to get rid of deleted recordings, tap this button, and then tap Immediately, After 1 Day, After 7 Days, After 30 Days (the default), or Never. If you implement a delay, you can recover deleted recordings from the Recently Deleted folder until the time period elapses.

» **Audio Quality:** Tap this button, and then tap Compressed (the default) or Lossless, as appropriate. Lossless gives full audio quality but takes more space.

» **Location-Based Naming:** Set this switch on (green) to have Voice Memos use the location as the root of the default name for new voice memos.

Putting Your Wallet on Your iPhone

The Wallet app enables you to store the details of your payment cards — credit cards, debit cards, and prepaid cards — on your iPhone and use them to make contactless payments via the Apple Pay service at either retail locations or online. You can also put other documents — such as boarding passes, movie tickets, and your driver's license — in Wallet for easy access. If your car supports digital keys, you can even add your car keys to Wallet and then use your iPhone or Apple Watch to lock, unlock, and start your car.

To get started with Wallet, tap its icon on the Home screen, and then tap the add icon (+) in the upper-right corner. On the Add to Wallet screen (see Figure 12-14), tap the type of item you want to add, and then follow the prompts. For example, tap the Debit or Credit Card button, and then use the Add Card screen to add the card's details either by scanning them with the camera or by typing them manually.

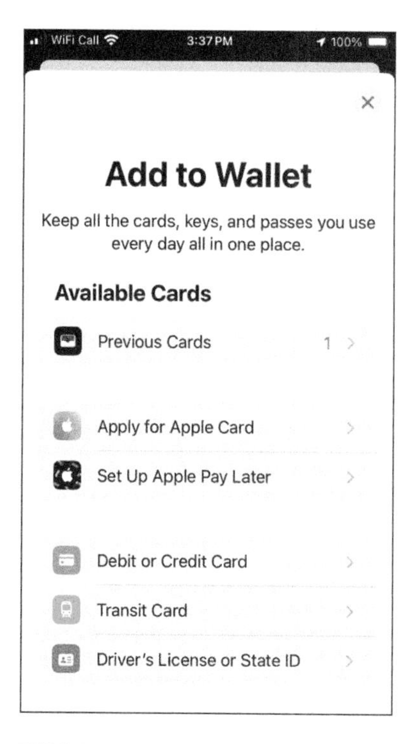

FIGURE 12-14: Choose which type of card to add to Wallet.

Sharing via AirDrop and NameDrop

Your iPhone includes two features for sharing items wirelessly with other Apple devices by using Bluetooth and Wi-Fi without an internet connection:

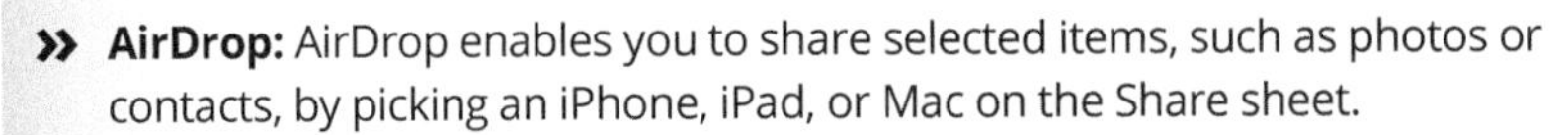

» **AirDrop:** AirDrop enables you to share selected items, such as photos or contacts, by picking an iPhone, iPad, or Mac on the Share sheet.

NEW

» **NameDrop:** New in iOS 17, NameDrop is a feature in AirDrop that lets you share items by bringing your iPhone, iPad, or Apple Watch close to another one of those three devices.

Configuring AirDrop and NameDrop

To make sure your iPhone is set up to use AirDrop and NameDrop, choose Settings➪General➪AirDrop. On the AirDrop screen, go to the Start Sharing By box, and then set the Bringing Devices Together switch on (green). While here, select the Receiving Off button or the Contacts Only button in the top box to set your iPhone's default AirDrop status. You can also choose Everyone for 10 Minutes to allow AirDrop with any available device for the next 10 minutes, but you would normally choose this setting via Control Center, as described next.

TIP

For security, choose the Receiving Off setting for AirDrop, and then enable receiving manually only when you need to receive items. To change your iPhone's AirDrop status quickly, open Control Center, and then long-press the Communications box (which contains the airplane mode, cellular data, Wi-Fi, and Bluetooth icons) until the box expands. Tap the AirDrop button, and then tap Receiving Off, Contacts Only, or Everyone for 10 Minutes, as needed.

Sending and receiving items via AirDrop

To send items via AirDrop, follow these steps:

1. **Open the appropriate app, and then select the items.**

   For example, open the Photos app, and then select one or more photos.

2. **Tap the share icon to display the Share sheet.**
3. **Tap AirDrop to display the AirDrop a Copy screen.**
4. **Tap the person or device to which to send the item or items.**
5. **When the *Sent* readout appears under the person or device, tap the Done button.**

Receiving items via AirDrop is equally easy. First, enable AirDrop either for Contacts Only or for Everyone for 10 Minutes, as explained in the preceding section. Then, when the AirDrop dialog opens, tap the Accept button (see Figure 12-15, left). The AirDrop dialog shows the transfer's progress (see Figure 12-15, middle). When the AirDrop Complete dialog appears (see Figure 12-15, right), tap it to display the received item in the appropriate app.

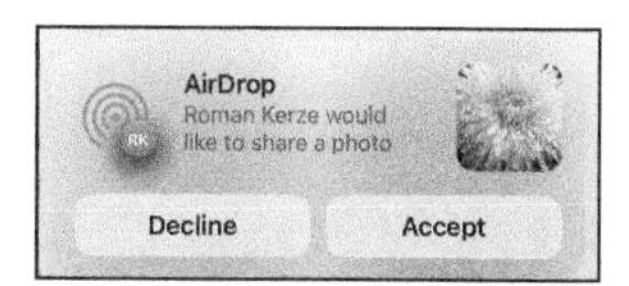

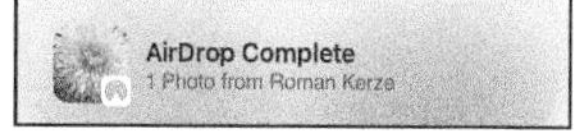

FIGURE 12-15: Receiving an item via AirDrop.

Sending and receiving items via NameDrop

NEW

To send an item via NameDrop, open the item, and then bring your iPhone close to another iPhone running iOS 17 or later or an iPad running iPadOS 17 or later. When the NameDrop screen appears (see Figure 12-16), tap the Share button.

When someone sends you an item via NameDrop, the item automatically opens in the appropriate app. For example, when someone sends you a photo, the Photos app opens and displays the photo.

FIGURE 12-16: Sharing a photo via NameDrop.

Controlling Lights, Locks, and More

The dedicated Home app is all about managing HomeKit accessories, which fit into a broader tech trend of connecting devices just about everywhere known as the Internet of Things (IoT). HomeKit is Apple's framework for such IoT accessories in and around the house — door locks, lightbulbs, security cameras, thermostats, garage-door openers, and so on. Inside Control Center, you can also glance at and control compatible accessories.

The Home app enables you to set up accessories in rooms or arrange for such accessories to come to life based on the time of day or when people come or go. For example, you can automatically arrange for the lights to come on when you arrive home from work.

In 2022, Apple joined other tech companies in supporting an emerging connectivity standard called Matter. The goal is to ensure that smart home accessories work well with each other, even across rival brands and platforms.

Connecting Your iPhone to Your Car

Your iPhone is almost always with you when you hop into your car. If your car supports Apple's CarPlay, you can connect your iPhone via either Bluetooth or a USB-C cable or Lightning cable. You can then display icons for the Phone, Music, Maps, Podcasts, and other apps, including a growing number from third-party publishers, right on your car radio or dash.

You can listen to songs, take or make calls, or get turn-by-turn directions through your car's speakers. You can interact with Siri, which acts as the voice of CarPlay. When you play music, you see the song title, artist, and other information on the car dash. When you're getting directions, you can display maps on the screen.

As of this writing, every major automotive manufacturer offers some CarPlay-capable vehicles, making for more than 600 models total. Some vehicles support wireless CarPlay; if yours is one of them, press the voice control button on your steering wheel to begin the setup. In some cars, you can use CarPlay to control the vehicle's radio and climate. In some new or recent-model cars that have more expansive screens, you can use widgets for apps such as Weather, Calendar, and Maps.

Your phone can help you remember where you parked, assuming you were using Bluetooth. Apple will record the location of the car and plot it on a map, which you'll find in today view. For backup, you can add notes or snap a photo when you park your vehicle.

Unlocking Your Car

If your car is compliant with Apple's car-key feature, you can add your car keys to the Wallet app, and then use your iPhone or your Apple Watch to unlock your car and to start it. You can share virtual keys with other people through the Messages app, giving those drivers full or restricted access. For example, if you're concerned about your teen driver, you can remotely lower the car stereo's volume and limit the car's top speed.

On compatible automobile models, you can even start the car with your phone and handle some of the functions of a key fob, including honking the horn, opening the trunk, and heating up the vehicle.

Measuring Things

The Measure app lets you measure objects by drawing virtual lines in free space. It also lets you check whether objects are level — for example, when you're hanging a picture on the wall and want to get it straight.

To use the Measure app, tap the Utilities folder on the Home screen, and then tap the Measure icon. Make sure the Measure tab at the bottom of the screen is active; if not, tap it. Then move the phone around so that a small white dot in a larger circle appears on your phone's display, at the edge of whatever it is you're measuring. Tap +, then drag the virtual line down to the end point of what you're measuring, and tap + again. You see the listed measurement on top of the virtual line you've drawn, shown either in the imperial or metric system (depending on your selection in Settings). You can also snap a picture of the virtual line or tap Clear to start over.

To check an object's level, tap the Level tab at the bottom of the screen, and then place your iPhone on the object. Adjust the object's level until the two circles overlap perfectly, the screen goes green, and the readout says 0°.

Translating Text or Speech

If you've ever been desperate to find a bathroom in a foreign country where you don't speak the language, or are merely a famished tourist seeking a decent place to eat, you'll surely appreciate the Translate app. As of this writing, the app can translate text or voice across more than a dozen languages, even when you're offline: English US and UK, Arabic, Chinese (Mandarin Simplified or Mandarin Traditional), Dutch, French, German, Indonesian, Italian, Japanese, Korean, Polish, Portuguese (Brazil), Russian, Spanish (Spain), Thai, Turkish, Ukrainian, and Vietnamese.

Tap the microphone icon and speak to have your words translated. Or type text using the keyboard native to the chosen language; iOS switches keyboards automatically when needed. You can save translations as Favorites for easy reuse. You can also hear a pronunciation read out loud.

To have a conversation, tap the Conversation tab. To specify how you and the other person are positioned, tap the View button, and then tap either Face to Face or Side by Side. In Face to Face view, the other person's controls and your translated text appear upside down at the top of the screen, allowing you to put your iPhone down on a table between you.

To translate signs or other written text, tap the Camera tab, and then select the From language and the To language in the pop-up menus. Point the camera at the text, and then tap the shutter button. Translate captures the text and displays a translation.

From any view, press the play arrow to hear the translated phrase spoken out loud.

Using the Health App

The Health app enables you to track a wide range of health data from various sources, including your Apple Watch (if you wear one), the Fitness app, third-party apps and devices (such as blood-pressure monitors), and data you input manually.

To get started with the Health app, tap the Health icon on the Home screen. The app opens, showing the Summary tab at first. There are two other tabs, the Sharing tab and the Browse tab, which I get to shortly.

Completing the Health Checklist and setting up your Medical ID

At the top of the Summary tab, you'll find the Health Checklist box. Tap the Review button to display the Health Checklist screen, and then work your way through its contents. You'll normally see the following:

» **Medical ID:** Your Medical ID gives information to first responders when they need it to assist you. If Medical ID is marked as Inactive, tap the Set Up button and follow the prompts to specify your information — your date of birth, medical conditions, medical notes, allergies and reactions, medications, weight, height, and so on. Designate one or more emergency contacts; set the Show When Locked switch on (green) to make your Medical ID accessible from the Lock screen; and set the Share During Emergency Call switch on (green) if you want to share your Medical ID when you call emergency services.

» **Emergency SOS:** Tap this button to display the Emergency SOS screen, and then set the Call with Hold and Release switch and the Call with 5 Button Presses switch on (green) or off (white), as needed.

» **Headphone Notifications:** Tap this button to set up iOS's monitoring of your headphone audio level.

» **Crash Detection:** Verify that Crash Detection is turned on. If not, tap the link to go to the Emergency SOS screen in the Settings app, and then set the Call After Severe Crash switch on (green).

» **Walking Steadiness Notifications:** If this item is marked as Inactive, tap the Set Up button, and then follow the prompts to enable the feature. You have to supply your date of birth, height, and weight for Health to calculate how wobbly you should be.

When you finish going through the Health Checklist items, tap the Done button.

Customizing the favorites on the Summary tab

Next, customize the favorites to make the Summary tab display the information you want, such as Steps, Blood Pressure, and Weight. Tap the Edit button in the upper-right corner to display the Edit Favorites screen, and then tap the Favorite star (filling it in blue) to the right of each item you want to display. The Existing Data tab shows a short list of items you can make favorites. Tap the All tab to display the full list. Tap the Done button when you finish customizing the list.

Setting up sharing on the Sharing tab

Next, tap the Sharing tab at the bottom of the Health screen, and then set up any sharing you want. You have three main options:

» **Share with Someone:** Tap this button to start sharing your health data with someone else.

- **Ask Someone to Share:** Tap this button to ask someone else to start sharing their health data with you.
- **Share with Your Doctor:** Tap this button to set up a secure connection to a health system so that you can share data with your doctor and download your health records.

Exploring health data on the Browse tab

Now tap the Browse tab at the bottom of the Health screen to display its contents. Here, you can explore the different categories of health data, such as Activity, Body Measurements, Heart, and Medications. Tap a category to see the various types of data it contains. For example, the Heart category contains AFib History, Blood Pressure, Cardio Fitness, Resting Heart Rate, and many more. Tap the item you want to view. On the resulting screen, you can tap the Add Data button to input data manually.

3

Creating and Enjoying Multimedia

IN THIS PART . . .

Enjoy music and podcasts on your iPhone.

Shoot photos with the Camera app, edit them, and share them.

Capture video with all the modes your iPhone offers, find videos, and watch them.

IN THIS CHAPTER

» Checking out your iPhone's Music app

» Bossing your tunes around

» Tailoring your audio experience

» Tapping into podcasts

Chapter 13
Enjoying Music and Podcasts

Your iPhone is great for working with audio and video. In this chapter, you learn to listen to music and other audio content on your iPhone. Chapter 15 covers video.

The iPhone's Music app is powerful but straightforward, so you'll be enjoying listening to your music in next to no time. You can customize your listening experience to make it even better. You can also enjoy podcasts using the Podcasts app, which can access huge numbers of podcasts on almost every topic under the sun and moon.

This chapter assumes you've already synced your iPhone with your computer or with iCloud, so your iPhone contains audio content, whether it be songs, podcasts, or audiobooks. If you haven't, turn back to Chapter 3 for instructions about syncing, or launch the iTunes Store app and buy a song so you'll have something to play.

Let's rock!

Introducing Your iPhone's Music App

To listen to music, launch the Music app by tapping the Music icon on the Home screen; this icon will be on the dock unless you've moved it. With the app open, you should see five icons at the bottom of the screen: Listen Now, Browse, Radio, Library, and Search. If you don't see these icons because the Now Playing screen is displayed, tap the handle at the top of the Now Playing screen to close it. If you're viewing the Account screen, tap the Done button to close it.

Before you dig into the Music app, you should know about the Apple Music and iTunes Match subscription services, because the Music app works a bit differently when you subscribe to one or both. Read the nearby "Apple Music and iTunes Match rock" sidebar to make sure you're up to speed on Apple Music and iTunes Match.

APPLE MUSIC AND ITUNES MATCH ROCK

iTunes Match and Apple Music are a pair of subscription music services offered by Apple. iTunes Match is the older of the two, designed to let you store all your music in iCloud so you can stream all your songs to any Mac, PC, or Apple device. It performs its magic by first determining which songs in your iTunes library are available in iCloud. And because tens of millions of songs are up there already, chances are that most of your music is already in iCloud. Then, iTunes proceeds to upload a copy of every song it *can't* match (which is much faster than uploading your entire Music library). The result is that you can stream any song in your iTunes library on any of your Macs, PCs, or Apple devices, regardless of whether the song files are available on the particular device.

As a bonus, all the music that iTunes matches plays back from iCloud is high quality — either 256 Kbps AAC or an even higher quality in the Apple Lossless Encoding format — even if your original copy was lower quality. You can even replace your lower-bit rate copies by downloading higher-quality versions.

You can store up to 100,000 songs in iCloud — songs you purchased from the iTunes Store don't count. The result is that only the tracks or albums that you specify are stored locally on your devices, saving gigabytes of precious storage space.

At just $24.99 a year, iTunes Match (`https://support.apple.com/en-us/HT204146`) is a bargain, but Apple Music, introduced in early 2015, may be a better option, albeit a more expensive one. For $10.99 a month (or $16.99 a month for you and up to five family members), your subscription provides access to more than 100 million songs (at last count) on demand. That's good, but what's even better is that Siri knows more about music than most people. Of course, Siri can play songs, albums, artists, or genres by name, but you can also ask Siri to play things such as the number-one song in October 1958 or music by a particular artist; in a few seconds, you'll be listening to whatever your heart desires. You can get Apple Music also as part of a subscription to Apple One, Apple's buy-in-bulk-and-save subscription service. See `https://support.apple.com/en-us/HT211659` for more information.

Both services require internet access, but as long as you're connected you can play any song from your iTunes library (iTunes Match) or any of more than 100 million songs (Apple Music) on your iPhone or other device signed in to your iCloud account. And since the music is streamed wirelessly, you don't have to worry about filling up your devices with music files.

Unless you have an unlimited data plan, you might want to disable iTunes Match and Apple Music over cellular by tapping Settings ➪ Music, and then going to the Allow Music to Access box and setting the Cellular Data switch off (white). Both subscription services are reasonably priced, but data surcharges could cost you a lot if you're not careful. And it's a good idea to have a backup of music you own before enabling either service, just in case.

Before you go off the grid — say, on a plane flight — tap the download button (a down-pointing arrow in a circle) for the song, album, or playlist you want to listen to at 30,000 feet. If you don't see a download button, tap the ellipsis (. . .) and choose Download.

It's music, just not your music

The Library tab gives you access to *your* music, where's the other four tabs — Listen Now, Browse, Radio, and Search — largely give you access to music that's *not* yours.

If you subscribe to Apple Music, you can listen to almost any song, album, or playlist you discover in any of these four tabs; you can also add these items to your library. If you don't subscribe, you're limited to songs you own and songs played on radio stations.

Here's what you'll find in each tab:

- **Listen Now:** Find music that Apple thinks you'll enjoy: playlists, albums, artists, and songs that (mostly) aren't already in your library.
- **Browse:** View the latest releases, plus song and album charts for dozens of musical genres as well as curated playlists featuring new songs and artists.

» **Radio:** Listen to online radio.

» **Search:** Search for pretty much any song, artist, album, or podcast.

The Listen Now tab and Browse tab operate in the same fashion: Each is a long page filled with content. Scroll down to see more; swipe left or right to see items, then tap any item to see additional information.

TIP

If you aren't an Apple Music subscriber and would prefer not to see the Listen Now tab and Browse tab (which sometimes feel like extended ads for Apple Music), choose Settings ⇨ Music, and then set the Show Apple Music switch off (white).

While you're exploring items on these tabs, tap the ellipsis (. . .) to see additional options, which may include Add to Library, Add to a Playlist, Play Next, Play Last, Share Song, and many more, depending on what is currently on-screen. Tap Love in this menu to tell the Music app you love this artist, song, album, or playlist, which helps the Music app suggest music it thinks you'll enjoy. Conversely, tap Suggest Less to tell the Music app you dislike this artist, song, album, or playlist, which also helps the Music app suggest music you'll enjoy.

Finally, if you're an Apple Music subscriber, tap the + button to add this song, album, or playlist to your library.

To start listening to radio, tap the Radio tab. At the top of the screen you'll find Music 1, Music Hits, and Music Country, Apple's trio of live radio stations on the air worldwide 24 hours a day, 7 days a week, and featuring world-class programming, interviews, and music. To listen to one, tap its play icon.

If you've listened to any of the radio stations, the next item on the Radio screen is the Recently Played section, which gives you instant access to stations you've listened to recently. Swipe right to left to see additional entries.

Further down the page, you'll find sections such as Discover New Shows, Artist Interviews, and Local Broadcasters. If nothing in these sections sounds good to you, scroll down to the bottom of the page and check out the last section, More to Explore, which provides a list of genres from Singer/Songwriter and Alternative & Indie to Rock. Tap a genre to see its stations; tap a station to begin listening.

Some radio stations, including Music 1, are available on your iPhone, iPad, Mac, PC, and Apple TV for free. You'll hear the occasional ad, or you can listen without ads if you subscribe to iTunes Match or Apple Music. If you subscribe to Apple Music, you also receive additional radio stations not available to non-subscribers.

When a station is playing, the name of the current song appears near the bottom of the screen (see Figure 13-1, left); tap it for additional options, as shown in Figure 13-1, right. You'll find these options for songs in your library and on Apple Music too.

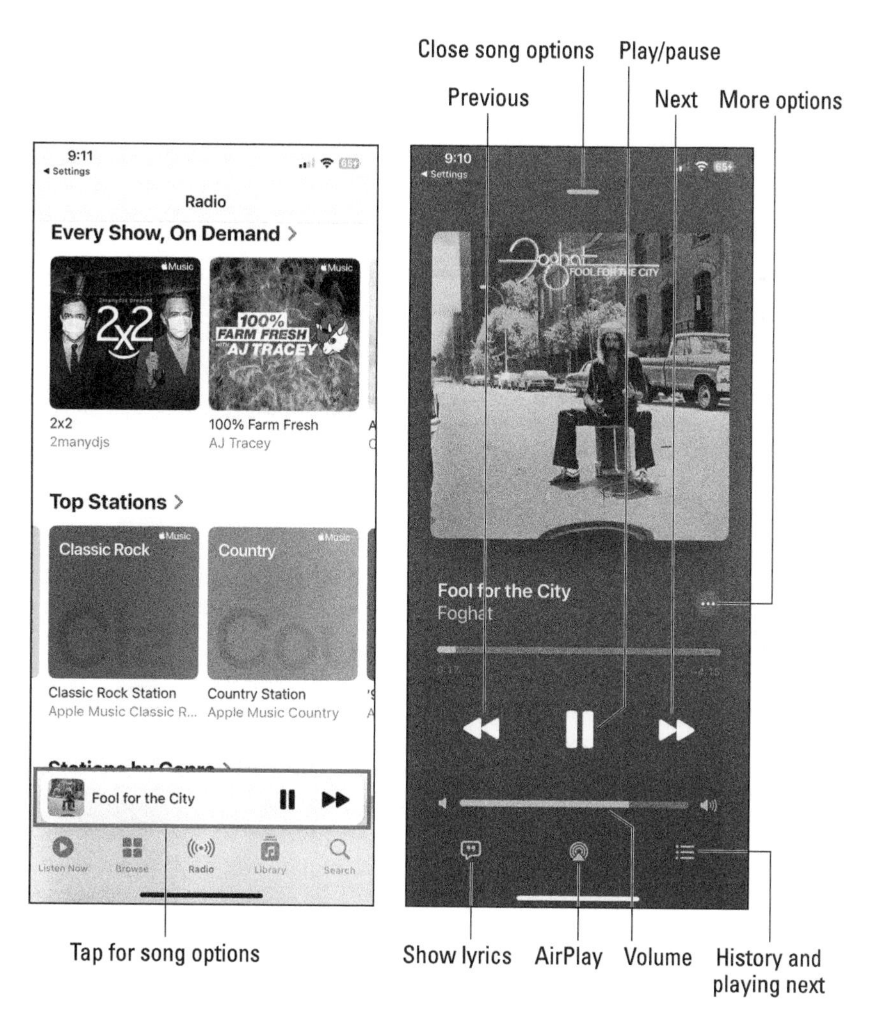

FIGURE 13-1: Tap the song's name (left) to see additional options (right).

Want stations to play more songs you'll like and fewer songs you dislike? When a song you love or hate is playing, tap the ellipsis (. . .) and then tap Love or Suggest Less. The more you do this, the better the suggestions in the Listen Now tab become.

My Music Is Your Music

To play music in your library, tap Library at the bottom of the Music screen. The library displays a list of the songs available to play, arranged in various categories. Tap Playlists, Artists, Albums, Songs, or Downloaded to see your music organized accordingly. Or tap Edit to enable additional criteria such as Music Videos, Genres, Compilations, and Composers. Further down the Library screen, you'll find the Recently Added section, which provides quick access to items you've added recently. The topmost items are the latest ones.

A library without library cards

REMEMBER

Music in the Library tab becomes available by syncing with iTunes or the Music app (as described in Chapter 3), subscribing to iTunes Match, adding music to your library from Apple Music, buying music in the iTunes Store, or any combination of the four. A music file doesn't have to be stored on your iPhone to be available on the Library tab. However, you must have internet access to play songs that are in your library but are stored in the cloud rather than on your phone. If you want to ensure that you can listen to a song, an album, or a playlist anytime you like, even without an internet connection, tap its ellipsis (. . .) and then tap Download. If your iPhone has only a modest amount of storage space, you'll need to strike a balance between downloading files that will occupy that storage and needing an internet connection to stream music from Apple's servers.

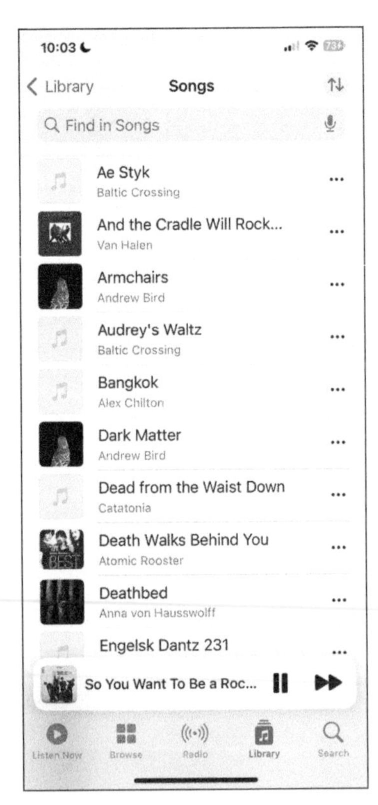

FIGURE 13-2: Tap a little letter on the right to see songs whose names begin with that letter.

You can browse to music by your preferred means. For example, you can tap Artists to display the list of artists, tap an artist to display their albums, and then tap an album to display its songs. You can then tap the song you want to play, or tap the Shuffle button at the top to play the songs in random order. Alternatively, you can simply tap Songs on the Library screen to display your songs in alphabetical order, as shown in Figure 13-2.

If the list is longer than one screen (which it probably is), flick upward to scroll down or flick downward to scroll up. Alternatively, tap one of the little letters on the right side of the screen to jump directly to songs whose names start with that letter. If you find it hard to tap accurately, drag your finger up and down the little letters to scroll to the right place.

Browsing your library is straightforward enough, but there's an even easier way to find an artist — or a song, an album, or a music compilation, for that matter. Tell Siri what you want to hear. All you have to do is ask (see "Using your voice to control your music," later in this chapter, for details).

If Siri isn't a viable option for you, you can achieve the same effect by typing your query. Tap Search in the lower-right corner and then tap in the search field at the top of the screen to activate the virtual keyboard. Now type the name of whatever you're looking for; tap the Apple Music tab or the Your Library tab, as appropriate; and then tap a search result.

Play it again, Liszt

Playlists let you create and organize collections of songs. They can evoke a particular theme or mood: opera arias, romantic ballads, British invasion, or whatever. Or you can make playlists of your favorite albums, favorite songs of all times, or greatest hits collections. (Younger folks sometimes call playlists *mixes;* older folks might think of them as the digital equivalent of a mix tape.)

Tap Playlists on the main library screen to view a list of your playlists. Tap a playlist and a list of the songs it contains appears. If the list is longer than one screen, flick upward to scroll down. Tap a song in the list and it plays. Or tap the Shuffle button at the top to start the playlist playing in random order.

TIP

It's easier to create playlists in the Music app (macOS Catalina or later) or in iTunes (macOS Mojave or earlier, or Windows) on your computer than on your iPhone.

As you see in Chapter 3, you can sync playlists with the Music or iTunes app on your computer. Or if you're an Apple Music subscriber, all your playlists are always available on all your devices.

Here's how to create playlists in the Music app on your iPhone:

1. **In the Music app, tap the Library icon at the bottom of the screen.**
2. **On the Library screen, tap Playlists to display the Playlists screen.**

3. **Tap New Playlist to open the New Playlist pane.**
4. **Tap the Playlist Title field and type a name for your new playlist.**
5. **If you want to add a description, tap the Description field and type a description of your new playlist.**
6. **If you want to add a picture, tap the camera icon in the picture at the top. Then tap Take Photo to snap a new photo or tap Choose Photo to browse for and select an existing photo.**
7. **Tap the Add Music button to display the Library screen.**
8. **To choose songs for this playlist, tap the means of browsing: Playlists, Artists, Albums, Songs, or whatever.**

   Or use the search field to find the song(s) you want to add.
9. **To add a song to your playlist, tap the + next to the song's name.**

   Repeat this process until you've added all the songs you want on this playlist.
10. **When you finish adding songs, tap the Done button in the upper-right corner.**

    This brings you back to the New Playlist pane, which shows all the songs you've added.
11. **Drag the songs into your preferred order by using the handles (three horizontal lines) at the right side of each song.**
12. **Tap the Done button in the upper-right corner to add the playlist to the Playlists screen.**

You want to play your new playlist right away? No problem! Tap the playlist's entry on the Playlists screen, and the playlist's songs appear. Tap the Play button to play the playlist from the beginning; tap a particular song to start there; or tap the Shuffle button to shuffle the playlist in exactly the way you couldn't with mix tapes.

If you create a playlist on your iPhone and then sync with your computer, that playlist remains on the iPhone and will also appear in the Music app or iTunes on your computer. Or if you're an Apple Music subscriber, the playlist will appear on all your other devices within a few minutes.

A playlist remain available on all devices until you delete it in the Music app or iTunes or on your iPhone. To remove a playlist in the Music app on the Mac or iTunes on the Mac or on a PC select the playlist's name in the source list and then press Delete or Backspace. To remove a playlist on your iPhone, long-press the playlist and then choose Delete from Library.

REMEMBER

You can edit playlists you created, but if you're an Apple Music subscriber, you can't edit playlists you've added from Apple Music.

To edit a playlist on your iPhone, tap the playlist you want to edit, tap the ellipsis (. . .) near the top of the screen on the right, and then choose Edit to do any (or all) of the following:

- **Move a song up or down in the playlist:** Drag the song up or down the playlist by using the handle (three horizontal lines) at the right side.
- **Add more songs to the playlist:** Tap the Add Music button at the top of the playlist.
- **Delete a song from the playlist:** Tap the – sign to the left of the song name, and then tap the Delete button that appears on the right. Deleting a song from the playlist doesn't remove the song from your iPhone or your library; it removes it only from this particular playlist.

When you finish editing, tap the Done button in the top-right corner of the screen.

Share and share alike

Home Sharing lets you use your iPhone to stream music, movies, TV shows, and other media content from your Mac or PC.

Then there's Family Sharing, which lets you share purchases and subscriptions from the iTunes Store, Book Store, and App Store with up to five other people in your family without sharing account information. Now you can pay for family purchases with the same credit card and approve kids' spending right from your iPhone or iPad. Here's the scoop on both types of sharing: Home and Family.

Sharing via the Home Sharing features

Turn on Home Sharing to stream music, movies, TV shows, and other media from your computer to your iPhone. Home Sharing is available only if your iPhone and your computer are on the same Wi-Fi network and use the same Apple ID.

To make Home Sharing work for you, first enable it on your computer, and then enable it on your iPhone. To set up Media Sharing on your computer:

- **macOS Ventura or later:** Launch System Settings, click General in the sidebar, and then click Sharing. Set the Media Sharing switch on (blue). In the Media Sharing dialog, select the Home Sharing check box, provide your Apple ID and password when prompted, and then click Done.

» **macOS Monterey or earlier:** Launch System Preferences, click Sharing, and then select the Media Sharing check box. Select the Home Sharing check box. The Enter the Apple ID Used to Create Your Home Share dialog opens. Verify the Apple ID, and then click Turn on Home Sharing, entering your Apple ID password when prompted.

» **Windows:** Open iTunes, choose Edit ➪ Preferences to open the Preferences window, and then click the Sharing tab at the top. Select the Share My Library on My Local Network check box to enable sharing. To specify what to share, select the Share Entire Library option button, or select the Share Selected Playlists option button and then select the check box for each playlist to share. Click OK to close the Preferences window.

From now on, as long as Music or iTunes is open, your library will be available for Home Sharing to all devices on your Wi-Fi network using the same Apple ID.

To stream media to your iPhone from your computer, tap Library at the bottom of the Music app's screen, and then tap Home Sharing. If shared libraries are available, you'll see them listed; tap one or more to view their contents on your iPhone for as long as you're connected to this Wi-Fi network.

If Home Sharing doesn't appear in the list of music sources, choose Settings ➪ Music, go to the Home Sharing area, and tap Sign In. In the dialog that opens, tap Sign In.

Now, when you look at your library, rather than just seeing the playlists, artists, and songs you've synced with your iPhone (and your Apple Music), you'll also see all the playlists, artists, songs, albums, and everything else in the iTunes (or Music) libraries on the Home Sharing computers as well. And you'll continue to see the shared content in your Music app as long as your iPhone remains connected to this Wi-Fi network.

To switch back to seeing just your music, just tap Library in the upper-left corner to return to your own playlists, artists, albums, and songs.

Movies and TV shows shared via Home Sharing don't appear in the Music app. Look for them in the TV app, as described in Chapter 15.

Sharing via the Family Sharing feature

Family Sharing lets up to six members of the same family share everything they buy at the iTunes, Book, and App Stores. It also provides sharing of family photos and a family calendar, location sharing, and more. If you have a family subscription to Apple Music, up to five additional family members can use Apple Music at no additional cost.

To set up Family Sharing, one adult in the household (known as the family organizer) invites up to five family members to join and agrees to pay for all iTunes, Book, and App Store purchases those family members make while part of Family Sharing.

When any family member buys an app, a song, an album, a movie, or an e-book, it is billed directly to the family organizer's account. The item is then added to the purchaser's account and shared with the rest of the family. The organizer can turn on Ask to Buy for any family member to require approval for any purchase. When a family member for whom Ask to Buy is active tries to buy an item, a notification is sent to the family organizer, who can review the item and approve or decline the request from their iPhone, iPad, or Mac.

The fine print says: You can be part of only one family group at a time and may switch to a different family group only twice per year. The features of Family Sharing may vary based on country and content eligibility.

To become the family organizer, open Settings on your iPhone, tap your name at the top of the screen, and then tap Set Up Family Sharing.

After you enable Family Sharing and invite your family members to join, everyone will have immediate access to everyone else's music, movies, TV shows, books, and apps, so you can download whatever you want to your iPhone with a tap anytime you like without having to share an Apple ID or password.

Taking Control of Your Tunes

Now that you have the basics down, take a look at some other things you can do with the Music app, starting with the controls you see when a song is playing.

REMEMBER

If you don't see the controls when a song is playing, tap the song name near the bottom of the screen.

Here's what you can do with the controls (see Figure 13-3):

» **Back icon:** Tap to return to whichever list you used last — Playlists, Artists, Songs, and so on. If you don't see this icon, you can go back to the previous screen by tapping the bar at the top of the playback screen or by swiping down on the playback screen.

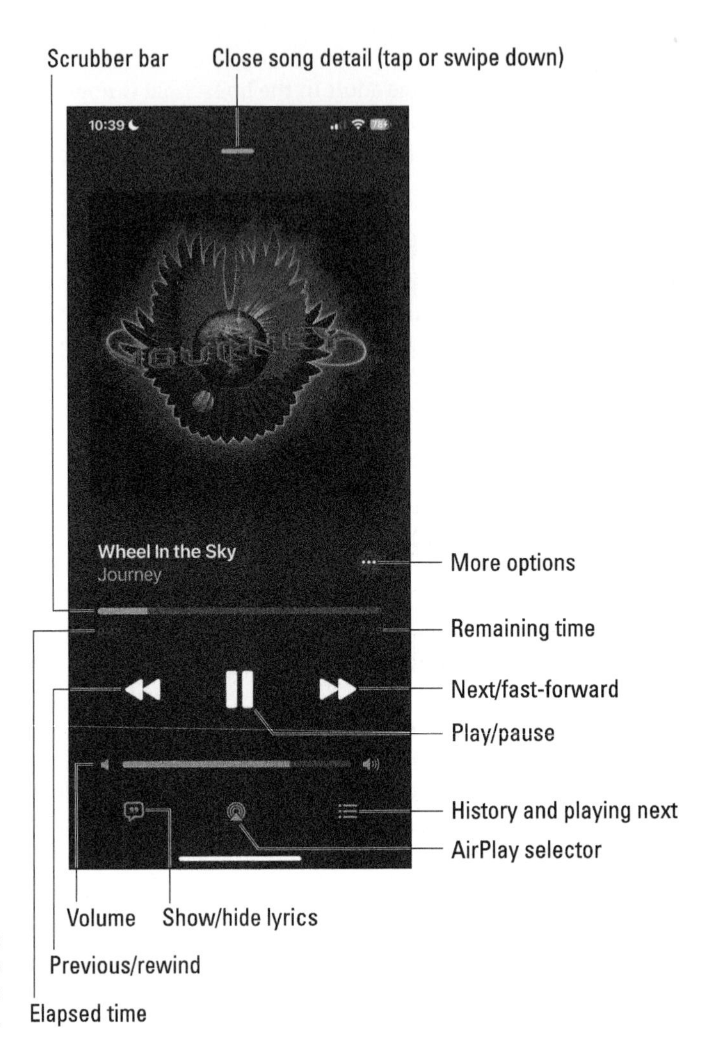

FIGURE 13-3: These controls appear when a song is playing.

- **Track position bar:** Tap anywhere in this bar and drag left or right to move the playhead through the song. The filled-in part of the bar shows how much of the song has been played. This bar is also called the *scrubber bar*, especially by those who speak of "scrubbing" backward or forward through songs.
- **Time elapsed and remaining:** Just below the track position bar on the left is the amount of time this song has played already (0:43 in Figure 13-3) and how much of this song remains to be played (3:29 in the figure).
- **Previous/rewind icon:** Tap once to go back to the beginning of the track. Tap again to go to the start of the preceding track in the list. Long-press this icon to rewind the song at double speed.

» **Play/pause icon:** Tap to play or pause the song.

» **Next/fast-forward icon:** Tap to skip to the next track in the list. Long-press this icon to fast-forward through the song at double speed.

» **Volume control:** Drag the end of the filled-in part of this bar to reduce or increase, respectively, the volume level.

» **Show/hide lyrics:** Tap this button to display the lyrics for the song. When the lyrics are displayed, see if the Apple Music Sing icon (shown in the margin) appears. (If the playback controls are hidden to give the lyrics room to express themselves, tap at the bottom of the screen to display the controls again.) If so, tap Apple Music Sing to reduce the vocal volume so you can sing the vocals. If you want to make finer adjustments to the volume, long-press the Apple Music Sing icon to display a larger volume control, and then slide that control up or down.

The Apple Music Sing icon appears only if the song has been configured so that the volume of the vocal track can be adjusted separately from the volume of the other tracks.

» **AirPlay selector icon:** You may or may not see the AirPlay selector icon (shown in the margin) on the playback screen (or in Control Center or elsewhere). (It's visible in Figure 13-3.) AirPlay lets you stream content wirelessly over Wi-Fi from your iPhone to any AirPlay-enabled device. Check out the "Playing with AirPlay" sidebar for the details.

» **Shuffle icon:** When viewing a list (songs, albums, and so on), tap once to shuffle songs and play them in random order. Tap it again to play songs in order again.

TIP

You can also shuffle tracks in any list of songs — such as playlists or albums — by tapping the shuffle icon, which appears at the top of the list.

» **Circled ellipsis (. . .) icon:** Tap this icon to download this song or album to your device, love it, suggest less like it, add it to a playlist, play it next, play it later, delete it from your library, or share it.

TIP

If you're using an Apple headset, you can squeeze the built-in control to pause, and squeeze it again to play. You can also squeeze it twice in rapid succession to skip to the next song.

You now know how to enjoy listening to music on your iPhone. But you can also customize your listening experience to make it even better. Read on, MacDuff!

PLAYING WITH AIRPLAY

AirPlay enables you to wirelessly stream music, photos, and video to AirPlay-enabled devices including (but not limited to) AppleTVs, SmartTVs, speakers, and receivers, and Macs running recent versions of macOS, such as macOS Sonoma.

The AirPlay Selector icon (shown here), which is available in some apps and Control Center, appears only if your iPhone detects an AirPlay-enabled device on the same Wi-Fi network. Tap the name of the AirPlay device you want to use, and whatever is playing on the Music app will be streamed to that device.

Customizing Your Audio Experience

To make your listening experience more enjoyable, you can normalize the playback volume and apply equalization. You can also set a sleep timer and use your voice to control your music.

Play all songs at the same volume level

The Sound Check option in the Music app and iTunes app automatically adjusts the level of songs so that they play at the same volume relative to each other. That way, one song never blasts out your ears even if the recording level is much louder than that of the song before or after it. To tell the iPhone to play songs at the same volume, tap Settings ➪ Music and set the Sound Check switch on (green).

TIP

You can also use Sound Check in Music or iTunes on your Mac or PC. On macOS Ventura or later, choose Music ➪ Settings; on macOS Monterey through Catalina, choose Music ➪ Preferences; or on earlier versions of macOS, choose iTunes ➪ Preferences. On Windows, choose Edit ➪ Preferences. Once the Settings dialog or Preferences dialog is open, click the Playback tab, select the Sound Check check box, and then click OK.

Choose an equalizer setting

An *equalizer* increases or decreases the relative levels of specific frequencies to enhance the sound you hear. Some equalizer settings emphasize the bass (low-end) notes in a song; other equalizer settings make the higher frequencies more apparent. The iPhone has more than a dozen equalizer presets, with names such as Acoustic, Bass Booster, Dance, Electronic, Pop, and Rock. Each is ostensibly tailored to a specific type of music or listening conditions (such as Small Speakers, which is not for stature-challenged orators).

To find out whether you prefer using equalization, listen to music while trying out different settings. To do that, first start listening to a song you like. **Then, while the song is playing, follow these steps:**

1. **On the Home screen, tap Settings ➪ Music ➪ EQ.**
2. **Tap different EQ presets (Pop, Rock, R&B, or Dance, for example), and listen carefully to the way they change how the song sounds.**
3. **When you find an equalizer preset that you think sounds good, leave it selected.**

If you don't like any of the presets, tap Off at the top of the EQ list to turn off the equalizer.

TIP

In Music on the Mac and iTunes on Mac and PC, you can set an equalization for an individual song. Locate the song, right-click it, and click Get Info to open the Info dialog for the song. Click the Options tab, pop open the Equalizer menu, and choose the equalization. Then click OK.

Setting a sleep timer

If you like to fall asleep with music playing but don't want to leave your iPhone playing music All Night Long (hat tip to Ritchie Blackmore and Rainbow), you can turn on its sleep timer. Here's how:

1. **On the Home screen, tap the Clock icon.**
2. **In the lower-right corner, tap the Timers icon.**
3. **Set the number of hours and minutes you want music to play, and then tap the When Timer Ends button.**
4. **Scroll down to the very bottom of the list and tap Stop Playing.**
5. **Tap the Set button in the upper-right corner.**
6. **Tap the Start button.**

If you have music playing already, you're finished. If not, launch the Music icon, and select the music you want to listen to as you fall asleep. When the specified time period elapses, the music stops playing and your iPhone goes to sleep.

Using your voice to control your music

Here's something cool: You can boss around your music by using nothing but your voice. Just press and hold down the Home or side button (or the equivalent button on a headset) and, after you hear the tone, you can:

- **Play an album, an artist, or a playlist:** Say "Play" and then say "album," "artist," or "playlist," and the name of the album, artist, or playlist, respectively. You can issue these voice commands at any time except when you're on a phone call or having a FaceTime video chat. You don't have to have music playing for these voice commands to work.
- **Shuffle the current playlist:** Say "Shuffle." This voice command works only if you're listening to a playlist or another list of songs, such as an album.
- **Find out more about the song that's playing:** Ask "What's playing?" "What song is this?" "Who sings this song?" or "Who is this song by?"
- **Get creative:** Siri just keeps getting better at playing music, so try asking it to play music with requests like the following:
  - Play songs popular in 2013.
  - Play music by Taylor Swift.
  - Play hit songs by David Bowie.
  - Play Chelsea Wolfe's first album.

Siri can usually interpret commands such as these, and within seconds you'll be listening to exactly what you want to hear. Plus, if you're an Apple Music subscriber, you can ask Siri to play virtually any song in the iTunes Store (100 million plus). Finally, because you're listening to this music on your iPhone, you won't look stupid when you're talking to it!

Controlling your music by speaking works most of the time, but in noisy environments Siri may mishear your verbal request and start playing the wrong song or artist or try to call someone on the phone. Using a headset helps. And syntax counts, so remember to use the exact wording in the list.

You can tap the words you just spoke and edit them with the keyboard if you like. And if Siri misspells a word, you can say, "that's spelled" and then speak the proper spelling letter-by-letter.

Siri can also identify songs playing in the background, using technology from the Shazam service. Just ask Siri, "What song is playing?" and let it listen for a moment. Siri will identify the song if it can.

Listening to Podcasts with the Podcasts App

A *podcast* is an on-demand radio program you listen to at your convenience by using the Podcasts app. To use the app, you must be signed in to your iTunes Store account. Normally, iOS keeps you signed in, but if you've signed out, choose Settings ⇨ Apple ID ⇨ Media & Purchases, and then tap Sign In. Type your username and password, and then tap OK.

When you're signed in to your iCloud account, tap the Podcasts icon on the Home screen to start finding and listening to podcasts.

The Podcasts app resembles the Music app with four icons at the bottom of the screen:

- **Listen Now:** Tap to see recent episodes of podcasts to which you subscribe and suggestions based on your listening habits. To manage notifications for your podcasts, tap your account icon (your monogram or your chosen image) in the upper-right corner, tap Notifications, and work on the Notifications screen.
- **Browse:** This icon leads to the podcasts store, which works like the Music or App stores with a few notable exceptions. First, all podcasts here are free. Second, you don't add or buy podcasts; you follow a podcast by tapping the +Follow button in the upper-right corner of its details page, or by tapping the ellipsis (. . .) even closer to the upper-right corner of the details page and then tapping Follow Show on the pop-up menu.

  Browse until you find a podcast that interests you, then tap it to see its details. From its details screen, you can listen to individual episodes or tap the + button to follow this show. If you follow a podcast, the Podcasts app will grab all new episodes as soon as they're released. After following a podcast, tap its ellipsis (. . .) icon for additional options, including Settings, Play Next, Share Show, and Unfollow Show.
- **Library:** The Library icon organizes your podcasts by Shows, Saved, Downloaded, and Latest Episodes. Tap one to see podcasts in your library that meet that criteria.
- **Search:** Type a word, phrase, podcast name, host name, or some other characteristic of what you seek, and the app displays podcasts that meet the criteria.

The Podcasts app works very much like the Music app. with minor exceptions. Like the Music app, you tap a podcast to listen to it, and tap its name at the bottom of the screen while it's playing to see its details, as shown in Figure 13-4.

Most of the controls on the details screen mimic the controls in the Music app, with the minor differences shown in Figure 13-4. The other exception is that when you swipe up on the details screen, instead of seeing lyrics above the Up Next section, you'll see Sleep Timer, Description, Chapters (if this podcast has chapters), and Episode Notes sections.

FIGURE 13-4: The Podcast app's controls are mostly the same as the Music app's controls.

Shopping with the iTunes Store App

The iTunes Store app lets you use your iPhone to download, buy, or rent a wide variety of content, including music, audiobooks, ringtones, and videos. And if you're fortunate enough to have an iTunes gift card or gift certificate in hand, you can redeem it directly from your iPhone in the iTunes Store app. Just sign in to your iTunes Store account as described in the preceding section, and then tap the iTunes Store icon on the Home screen to start shopping.

IN THIS CHAPTER

» Taking photos

» Choosing settings for the Camera app

» Viewing and organizing your photos

» Editing your photos

» Sharing your photos

Chapter 14

Shooting, Editing, and Sharing Photos

Your iPhone's cameras are among the best smartphone cameras available. The front (screen side) camera is optimized for selfies and FaceTime calls, while the main camera on the rear of the iPhone is designed to take a wide range of photos and videos. Depending on the iPhone model, the rear camera may have a single lens or multiple lenses, such as an ultrawide lens, a main lens, and one or more telephoto lenses.

This chapter covers taking photos with the Camera app. The next chapter covers shooting videos.

To help you organize your photos and make the most of them, your iPhone includes the Photos app. Photos enables you to view and edit your photos, organize them into albums, and share them easily with other people.

Taking Photos with the Camera App

The Camera app makes taking photos and videos as easy as possible — but it also enables you to choose settings manually when you need extra control. This section shows you how take photos. The next chapter explains how to shoot videos.

Opening the Camera app

To get started with the Camera app, open the app in one of these ways:

- **Home screen:** Tap the Camera icon. To open the app in a particular mode, long-press the Camera icon, and then tap Selfie, Video, Portrait, or Portrait Selfie on the pop-up menu.
- **Lock screen:** Tap the Camera icon. Alternatively, swipe left on the Lock screen. This move is useful when it's hard to see the Camera icon, such as in bright sunlight.
- **Control Center:** Tap the Camera icon. Or, to open the app in a particular mode, long-press the icon, and then tap Selfie, Video, Portrait, or Portrait Selfie on the pop-up menu.
- **Siri:** Say something like "Hey Siri, open the Camera app."
- **Apple Watch:** Open the Camera Remote app.

FIGURE 14-1: The Camera app gives you instant access to its key controls.

Meeting the Camera app's controls

Once you've opened the Camera app, you should see controls similar to those shown in Figure 14-1. The controls vary somewhat depending on which iPhone model you're using, so don't worry if things look a bit different. The subject should be different too.

These are the main controls, in roughly descending order of importance:

- **Camera mode:** This bar shows the current camera mode, such as photo mode, and enables you to switch to other modes, such as video mode, slo-mo mode, or pano (panorama) mode. Depending on the iPhone model, not all the camera modes may fit on the bar, so you may need to slide it left or right to reach the mode you want.
- **Camera chooser:** Tap this icon to toggle between the rear camera and the front camera.
- **Lenses:** If your iPhone's rear camera has multiple lenses, tap the button for the lens you want to use.
- **Take picture:** Tap the big white button to take a picture using the current camera mode.
- **Photo and video viewer:** Tap this thumbnail, which shows the last photo or video, to view that photo or video in the Photos app.
- **Flash:** Tap the lightning bolt icon to toggle the flash between the auto setting and the off setting. If the flash is set to on, which you learn how to do shortly, tapping this icon once sets the flash to off, and tapping it again sets the flash to auto rather than setting it back to on.
- **Expand:** Tap the arrowhead icon to expand the camera controls bar, which is hidden by default. When displayed, the camera controls bar replaces the camera mode bar. See the section "Using the camera controls bar," later in this chapter.
- **Live Photo:** Tap this icon to toggle the Live Photo feature on or off. See the section "Shooting and enjoying live photos," later in this chapter, for coverage of live photos.
- **Camera active indicator:** This green dot shows that the camera is active.

Taking a photo and viewing it

To get the hang of the Camera app, try taking a photo and then viewing it in the Photos app. Follow these steps:

1. **Open the Camera app using one of the methods explained in the previous section. For example, tap the Camera icon on the Home screen.**

   The screen shows what the camera's lens is viewing.

2. **If you need to switch from the rear camera to the front camera, tap the camera chooser icon.**

3. **If necessary, change the camera mode by tapping the appropriate mode on the camera mode bar.**

   For example, tap Portrait if you want to take a portrait.

   This example uses photo mode, which is normally the default mode for the Camera app.

4. **Aim the lens at your subject, and use the photo frame to compose the shot.**

5. **If you need to switch to a different lens, tap its button.**

   For example, tap the 2x button to switch to the telephoto lens.

6. **Tap the take picture button or press either the volume up button or the volume down button.**

   The Camera app takes the photo, and its thumbnail appears in the photo and video viewer.

7. **Tap the photo and video viewer to display the photo in the Photos app (see Figure 14-2).**

   You can then work with the controls explained in the following list. Tap the back icon (<) when you're ready to return to the Camera app to take more photos.

FIGURE 14-2: Viewing your latest photo in the Photos app.

Here's what the controls in the Photos app do:

- **Back:** Tap the < icon to return to the Camera app.
- **Live:** This icon and the *LIVE* readout indicate that the photo is a live photo, one that includes video. See the next section for more information.
- **Photo date and time:** This readout shows the date and time the photo was taken.

» **All Photos:** Tap this button to go to the All Photos album.

» **Actions:** Tap the three dots icon to display the Actions menu, which includes commands such as Duplicate, Hide, Adjust Date & Time, and Adjust Location.

» **Photo chooser:** Tap the photo you want to display. Scroll left or right as needed.

» **Share:** Tap the arrow-escaping-a-box icon to display the share sheet. See the section "Sharing Your Photos," later in this chapter.

» **Favorite:** Tap the heart icon to bestow or remove favorite status.

» **Info:** Tap the circled-*i* icon to display the info pane for the photo.

» **Edit:** Tap this button to open the photo for editing. See the section "Editing Your Photos," later in this chapter.

» **Delete:** Tap the trashcan icon to start deleting the photo, and then tap the Delete Photo button in the confirmation dialog.

You now know the basic moves for taking a photo and viewing it. Next, let's look at other techniques you can use when taking photos.

Shooting and enjoying live photos

A *Live Photo* is one that includes three seconds of video, half of it before the still frame and the other half after the still frame. A Live Photo plays automatically when you display it in the Photos app; you can make it play again by long-pressing the photo.

Perhaps the best thing about Live Photos is that they enable you to change the key frame, the still photo. Tap the Edit button to open the photo for editing, and then tap the Live tab at the bottom of the screen. You see a frame-by-frame scrubber bar below the picture, with a white rectangle marking the key frame. Drag this white rectangle to the frame you want to use, lift your finger, and then tap the Make Key Photo button that appears. You can also shorten the video clips by dragging the start marker and the end marker. Tap the Done button when you're satisfied with your changes.

TIP

Live Photos can be entertaining, but to be ready to shoot a Live Photo, the Camera app needs to record video through the active camera constantly while the app is active. The app can then pull the 1.5 seconds of video before you took the picture from its buffer to create the first part of the live photo. Because only 1.5 seconds are needed, the footage doesn't need much storage — the Camera app can just keep deleting any footage more than a couple of seconds old. But shooting video does take up battery power, so turn off Live Photos if you don't use them.

Tap the LIVE pop-up menu in the upper-left corner of a Live Photo to display five options:

- **Live:** This is the standard effect. It's selected until you change it.
- **Loop:** This effect plays the Live Photo continuously.
- **Bounce:** This effect plays the live action forward, and then reverses it. This is sometimes fun (people walking backwards) and sometimes less so (your child unswallowing food).
- **Long Exposure:** This effect blurs moving people or objects, leaving stationary people and objects sharp.
- **Live Off:** You may be tempted to add "the land" or "your wits," but this effect turns off playing the live footage for the photo.

TIP

You can create a still version of a Live Photo by displaying that photo, tapping the actions icon (three dots), and tapping Duplicate on the menu that opens. Then, in the dialog that opens at the bottom of the screen, tap Duplicate as Still Photo.

Using the camera controls bar

When you need to exercise more control over a shot than you get from the controls the Camera app displays by default, tap the expand icon at the top of the screen. The collapse icon replaces the expand icon and the camera controls bar appears in place of the camera mode bar.

The selection of controls on the camera controls bar varies depending on the iPhone model, because the camera modules on different iPhone models have different capabilities. On some models, the controls fit within the screen width, whereas on other models, you need to scroll the camera controls bar to reach some of the controls. The left and right screens in Figure 14-3 show a fully-stocked camera controls bar with its controls labeled. When you tap one of the controls, its settings buttons appear on the camera controls bar so that you can pick the setting you want. Once you've done that, tap the control's button again to display the top level of the camera controls bar.

Here's what the controls on the camera controls bar do:

- **Flash:** Tap the lightning bolt icon to display the flash controls, and then tap the Flash Auto button, the On button, or the Off button, as needed.
- **Night mode:** Tap this icon to display an exposure control slider for taking photos in low light. The night mode feature is also called low light.

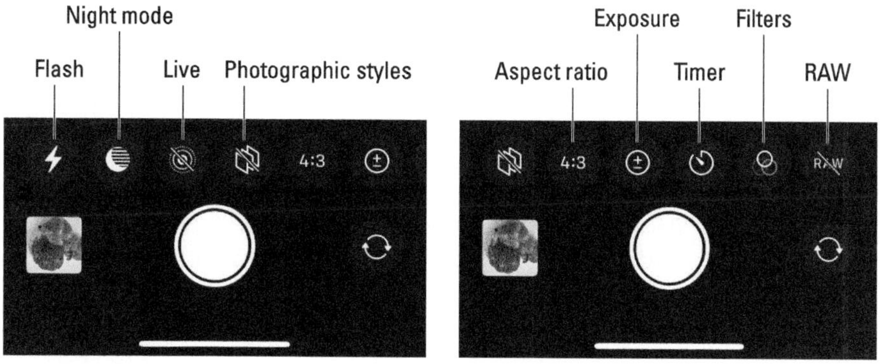

FIGURE 14-3: The controls on the camera controls bar are tailored to the iPhone model's capabilities.

- **Live:** Tap this icon to display the Live controls, and then tap the Auto button, the On button, or the Live Off (your savings) button.
- **Photographic styles:** Tap this icon to display the photographic styles controls, and then swipe left one or more times until you reach the style you want to use.
- **Aspect ratio:** Tap this icon to display the aspect ratio controls, and then tap the 4:3 button, the Square button, or the 16:9 button, as needed.
- **Exposure:** Tap this icon to display the exposure slider, and then drag the slider to set the exposure you want. Values range from –2.0 stops to +2.0 stops.
- **Timer:** Tap this icon to display the Timer controls, and then tap the Timer Off button, the 3s button, or the 10s button.
- **Filters:** Tap this icon to display the filter chooser, and then tap the filter you want to apply.
- **RAW:** Tap this icon to display the RAW controls, and then tap the On button or the RAW Off button. RAW is available only on iPhone 12 Pro and later Pro models. If this icon doesn't appear, choose Settings ⇨ Camera ⇨ Formats, and then set the Apple ProRAW switch on (green).

The following subsections tell you how to put these features to use. I skip live photos, because you've met them already, but you'll add zooming with the zoom wheel.

Zooming with the zoom wheel

On an iPhone that has multiple lenses, you can zoom by using the zoom wheel feature as well as by pinching and unpinching. To use the zoom wheel feature, long-press any of the lens buttons (such as 2x), and then rotate the wheel that appears (see Figure 14-4).

Taking photos with flash

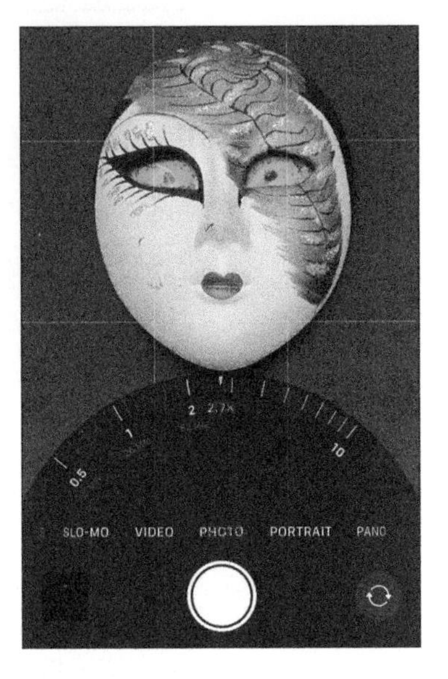

FIGURE 14-4:
Long-press any lens button to display the zoom wheel, and then rotate it to zoom.

The rear camera includes an LED (light-emitting diode) flash to add light when your photos need it. Tap the Flash icon in the upper-left corner of the screen to toggle the flash between the auto setting and the off setting. Or, to turn the flash on all the time, tap the expand icon at the top of the screen, tap the flash icon on the camera controls bar, and then tap On.

Recent iPhones include what Apple calls True Tone Flash with Slow Sync. This feature slows the shutter speed when you shoot photos using flash, thus letting in more light to give a better exposure of your subject and its background.

The front camera uses the iPhone's screen as its flash. The camera makes the screen brighten with TrueTone lighting for just a moment, enough to light those selfies you take in dim lighting.

TIP

When shooting a subject with strong light behind it, turn the flash on to provide fill-in lighting.

Using night mode

The iPhone 11, 12, 13, 14, and 15 models include the night mode feature, which enables you to shoot in dim light without using the flash. The results are often better than using the flash — provided you can hold the iPhone steady. Mounting your iPhone on a tripod or another stabilizing device helps hugely.

The Camera app automatically turns on night mode when conditions warrant it. The night mode icon appears in the upper-left corner of the screen, showing the exposure time, such as 3s in the left screen in Figure 14-5. If you need to adjust the exposure manually, tap the expand icon at the top of the screen, tap the night mode icon on the camera controls bar, and then drag the vertical white line along the exposure slider (see the right screen in Figure 14-5). Tap the take picture button when you're ready to capture the photo. The exposure slider shows a countdown of the time remaining, letting you know how much longer you need to hold the iPhone steady.

In Figure 14-5, the photo preview looks iffy at best. But if you look ahead to Figure 14-12, you can see that the photo came out pretty well.

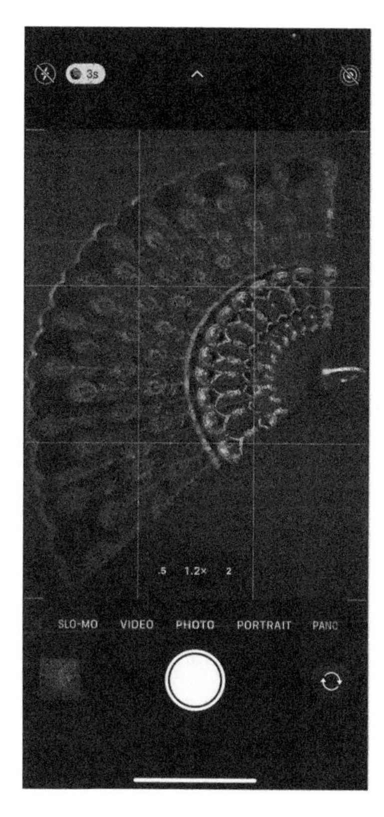

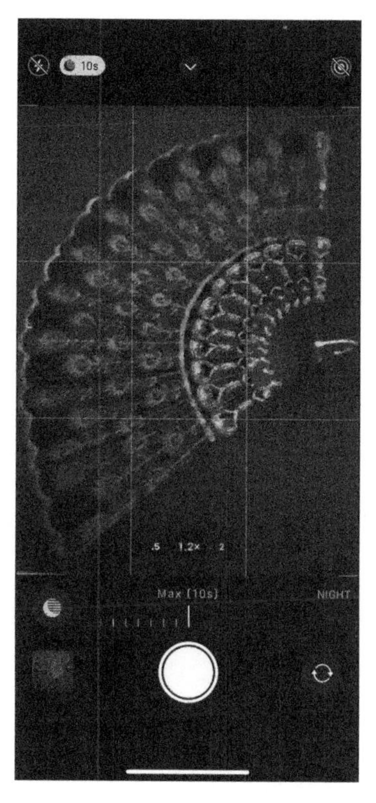

FIGURE 14-5: Night mode enables you to shoot in dim light without using the flash.

Using photographic styles

If your iPhone includes the photographic styles feature, you can switch styles by tapping the expand icon to display the camera controls bar, tapping the photographic styles icon, and then swiping left or right to select the style you want. The Camera app continues using that style until you change to another style — so if you're changing styles for just a few photos, remember to change back.

Shooting photos with different aspect ratios

The Camera app enables you to shoot photos with different aspect ratios: 4:3 (the default), square (1:1, if you prefer numbers), or 16:9. To change the aspect ratio, tap the expand icon to display the camera controls bar, tap the aspect ratio icon, and then tap the aspect ratio you want. The Camera app continues using this aspect ratio until you change it again or until your current Camera session ends.

TIP

To make the aspect ratio stick between sessions in the Camera app, choose Settings ⇨ Camera ⇨ Preserve Settings, and then set the Creative Controls switch on (green). This switch affects the filter (discussed in the upcoming "Applying filters to your photos" section) as well as the aspect ratio setting.

Setting the exposure manually

To set the exposure manually, tap the expand icon at the top of the screen to display the camera controls bar, and then tap the exposure icon. Slide the exposure slider to the right to reduce the exposure by up to 2.0 stops, or slide it to the left to increase the exposure by up to 2.0 stops. The exposure indicator appears in the upper-left corner of the screen, next to the flash indicator, as a reminder that you have applied an exposure adjustment (see Figure 14-6).

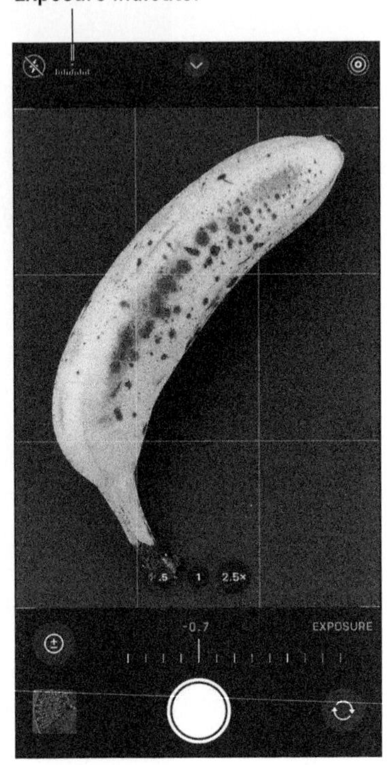

FIGURE 14-6: Use the exposure slider above the take picture button to adjust the exposure.

The Camera app continues using the exposure adjustment until you change it or until the end of your session in the Camera app; if you want your chosen exposure to stick beyond that, choose Settings ⇨ Camera ⇨ Preserve Settings, and then set the Exposure Adjustment switch on (green).

Using the timer

To take a photo with the timer, tap the expand icon at the top of the screen to display the camera controls bar, tap the timer icon, and then tap the 3s button or the 10s button. A timer icon appears in the upper-right corner of the screen showing the delay. Tap the take picture button to start the countdown.

If the Live Photos feature is enabled, the Camera app takes a single live photo. If Live Photos is disabled, the Camera app takes a burst of 10 photos to increase your chances of getting a good shot.

TIP

If you wear an Apple Watch paired with your iPhone, you can use the Camera Remote app on the Apple Watch to take photos on the iPhone. The Camera Remote app includes a 3-second timer that's great for taking selfies or group photos with the iPhone mounted on a tripod.

Applying filters to your photos

You can give a photo a different look by applying a filter to it. The Camera app enables you to apply a filter before taking a photo, and the Photos app lets you apply a filter afterward. You can also remove a filter at any time.

To apply a filter before taking a photo, tap the expand icon at the top of the screen to display the camera controls bar, tap the filters icon to display the filter chooser, and then tap the filter you want, such as Vivid Warm, Dramatic, or Noir. Then take the photo as usual.

To apply a filter after taking a photo, tap the filters icon to display the filter chooser, and then tap the filter. To remove the current filter, tap the Original filter at the left end of the filter chooser.

Taking RAW photos

For most photos, the Camera app processes the image data to adjust the exposure, improve the colors, and remove noise (inaccurate data) and artifacts (such as curvature from the wide-angle lens). The Camera app also compresses the image data and saves it in the HEIF format or the JPEG format. This approach gives good results at an acceptable file size, but some image data is inevitably and irrevocably lost.

If you need full image quality, and your iPhone supports the RAW format, you can shoot RAW photos instead. RAW photos are full-quality files that contain uncompressed and unprocessed image data straight from the camera sensor. They will need post-processing to give a polished result.

To shoot RAW files, see if the RAW button appears in the upper-right corner of the screen. If it doesn't, choose Settings ⇨ Camera ⇨ Formats, and then set the Apple ProRAW switch on (green). Now tap the RAW button, so that it shows *RAW* without a line through it. Alternatively, tap the expand icon at the top of the screen to display the camera controls bar, tap the RAW button, and then tap the On button.

The Camera app maintains your RAW setting until the end of the Camera session. To preserve the RAW setting between sessions, choose Settings ⇨ Camera ⇨ Preserve Settings, and then set the Apple ProRAW switch on (green).

The iPhone 15 Pro models can transfer ProRAW photos directly to a Mac via a USB-C cable. This capability enables you to use an iPhone 15 Pro as your primary camera.

Shooting portraits

If the Camera app on your iPhone includes portrait mode on the camera mode bar, you can use this mode to isolate a photo's subject from the background and apply assorted lighting effects that can hugely change (and often improve) the photo.

Start by lining up your subject, as in the left screen in Figure 14-7. Here, the background is fully visible, including a bookshelf.

FIGURE 14-7: Line up your subject (left), and then tap the Portrait button to switch to portrait mode. The background blurs.

Now tap the Portrait button on the camera mode bar. As you can see in the right screen in Figure 14-7, the Camera app blurs the background and applies the Natural Light lighting to the subject.

Long-press the Natural Light button to display the lighting wheel (see the left screen in Figure 14-8), and then rotate the wheel counterclockwise to try the different lighting styles: Studio Light, Contour Light, Stage Light, Stage Light Mono, or High-Key Light Mono. When you find the lighting style you want, release the wheel. You can then shoot the portrait (see the right screen in Figure 14-8).

Shooting bursts of photos

To improve your chances of capturing the action or the expression you're after, you can shoot bursts of photos. To shoot a burst, drag the take picture button toward the photo and video viewer. So if you're holding the iPhone in portrait orientation, drag the take picture button to the left; if you're holding the iPhone in landscape orientation with the take picture button on the left, drag it up; and for landscape orientation with the take picture button on the right, drag it down. The Camera app displays the number of photos in the burst (see the left screen in Figure 14-9) and plays a rapid shutter sound to encourage you. Release the take picture button to end the burst.

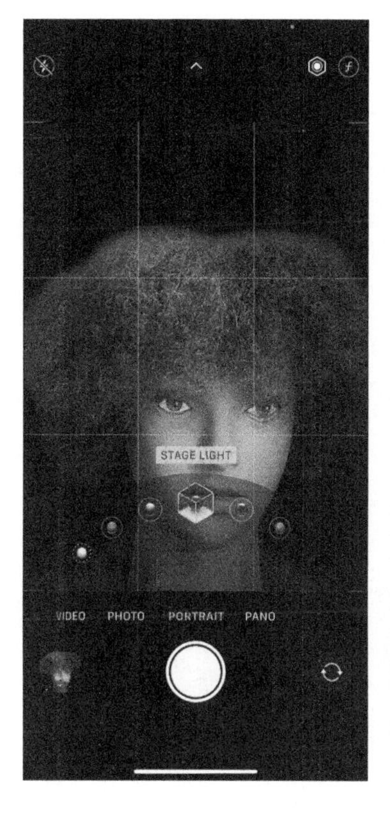

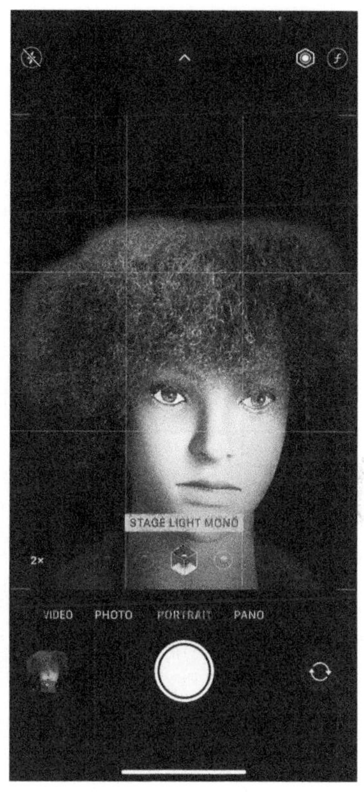

FIGURE 14-8: Rotate the lighting wheel to try different lighting styles (left). Release the wheel when you're ready to take the portrait (right).

TIP

You can also take a burst by holding down the volume up button. To shoot bursts this way, choose Settings ⇨ Camera, and then set the Use Volume Up for Burst switch on (green).

Once you've shot a burst, you can open it and decide which shots to keep. To get to the burst, either tap the photo and video viewer in the Camera app straight after shooting the burst, or go to the Photos app, tap the Bursts album, and then tap the burst. You'll notice that the thumbnail for the burst looks like it's sitting on a stack of photos. The thumbnail is the picture that the Camera app decided unilaterally best represents the burst. (You may disagree.)

As you can see in the right screen in Figure 14-9, the word *Burst* and the number of photos appears in the upper-left corner of the burst's screen. Tap the Select button at the bottom of the screen to display the contents of the burst. The selected image from the burst appears front and center, bordered by the edges of other photos in the burst.

FIGURE 14-9: Drag the take picture button to the left to shoot a burst (left). In the Photos app, the Burst icon and readout in the upper left indicate a burst shot.

At the bottom of the display is the photo chooser, a strip of thumbnails representing the pictures in the burst. Below one or more of these images is a gray dot, indicating that the photo is one the Camera app reckons is the best or among the best in the burst. Scroll left or right to examine the other pictures in the burst. A gray triangle above the thumbnails shows you your location.

When you identify an image you want to keep, tap the selection circle in its lower-right corner, placing a check mark on it. When you've selected all the images you want, tap the Done button. A dialog opens, asking if you want to keep the other photos in the burst. Tap the Keep Everything button or the Keep Only *n* Favorites button, as appropriate.

Shooting panoramas

To shoot a panorama, tap the Pano button on the camera mode bar. Holding the iPhone in portrait orientation, aim the lens at your start point, and then tap the take picture button. Rotate the iPhone, and perhaps yourself, clockwise, keeping the white arrow on the horizontal guide line. Tap the white square-in-a-circle stop icon to stop shooting the panorama.

Choosing Settings for the Camera App

To get the most out of the Camera app, spend a few minutes configuring the app's settings. To get started, tap the Settings app icon on the Home screen, and then tap the Camera button on the Settings screen. The Camera screen appears (see Figure 14-10, left and right).

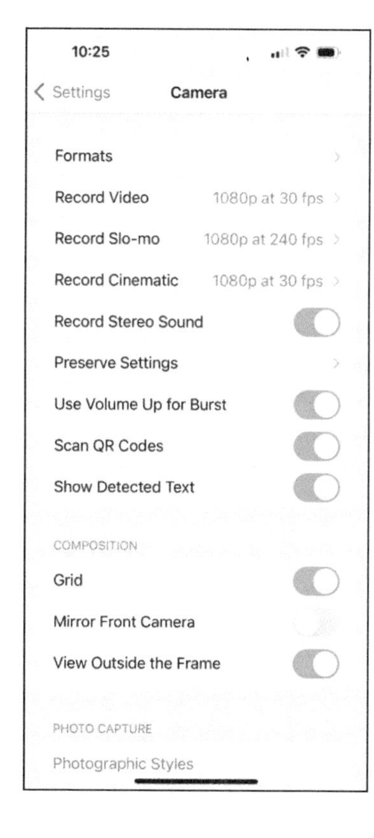

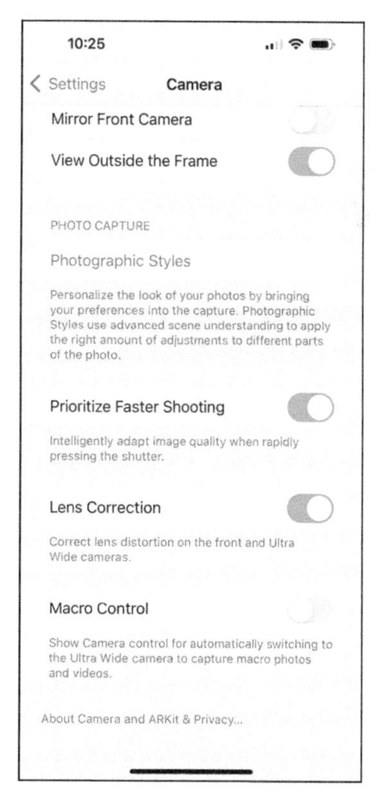

FIGURE 14-10: Work on the Camera screen in the Settings app (left and right) to configure the Camera app.

REMEMBER

The controls available on the Camera screen in Settings vary depending on the capabilities of your iPhone's cameras, so don't worry if you don't see some of the controls listed here.

The following list explains the settings you can configure in the unnamed box at the top of the Camera screen:

- **Formats:** Tap this button to display the Formats screen. In the Camera Capture box, tap the High Efficiency button or the Most Compatible button, as needed. Choosing the High Efficiency setting makes the Camera app use the High

Efficiency Image Format (HEIF) for photos and the High Efficiency Video Coding (HEVC) format for videos, giving high quality with the most compact file size possible. Choosing the Most Compatible setting makes the Camera app use the widely supported JPEG format for photos and the very widely used H.264 format for most videos. However, videos you shoot at 4K resolution at 60 frames per second (fps) and 1080p resolution at 240 fps still use the HEVC format.

- **Record Video:** Tap this button to display the Record Video screen, and then choose settings for shooting video. See the next chapter for advice.
- **Record Slo-Mo:** Tap this button to display the Record Slo-Mo screen, and then choose settings for shooting slow-motion video. Again, see the next chapter.
- **Record Cinematic:** Tap this button to display the Record Cinematic screen, and then choose settings for shooting Cinematic video. For this, too, see the next chapter.
- **Record Stereo Sound:** Set this switch on (green) to record in stereo rather than mono.
- **Preserve Settings:** Tap this button to display the Preserve Formats screen, which provides various switches for preserving settings from one session in the Camera app to the next rather than resetting them to their defaults. Set the Camera Mode switch, the Creative Controls switch, the Depth Control switch, the Exposure Adjustment switch, the Night Mode switch, the Portrait Zoom switch, and the Live Photo switch on (green) or off (white) to tell which settings to preserve. For example, if you typically shoot video rather than still photos, set the Camera Mode switch on (green). Then, any time you launch the Camera app, it starts in the mode you were last using, such as video mode.
- **Use Volume Up for Burst:** Set this switch on (green) to be able to shoot a burst of photos by pressing and holding the volume up button.
- **Scan QR Codes:** Set this switch on (green) to have the Camera app automatically scan any quick response (QR) codes it identifies in the frame.
- **Show Detected Text:** Set this switch on (green) to have the Camera app display any text it has identified in the frame.

In the Composition box, you can configure these settings:

- **Grid:** Set this switch on (green) to display a nine-rectangle grid on-screen to help you compose shots.

- **Level:** Set this switch on (green) to display a three-part horizontal line across the middle of the screen when the iPhone is tilted. The central section of the line shows the degree of tilt, letting you level the iPhone by aligning the ends of the central section with each outside section.
- **Mirror Front Camera:** Set this switch on (green) to save photos taken with the front camera in the same orientation as the camera preview, which flips them horizontally to give you the effect of looking in a mirror. Set this switch off (white) to save photos unflipped, as another person would see you.
- **View Outside the Frame:** On an iPhone that includes the ultrawide lens, set this switch on (green) to have the preview display areas outside the frame of the photo. This extra data comes from the ultrawide lens and shows what you could include in the photo by switching to that lens.

In the Photo Capture section of the Camera screen, you can configure the following setting:

- **Photographic Styles:** If this button appears, tap it to display the Photographic Styles screen, which enables you to apply your choice of photographic style to all the regular photos you take. The Camera app provides built-in photographic styles, including the Standard style (which gives a true-to-life look and used as the default), the Rich Contrast style (which reduces tone to give a more dramatic look), and the Vibrant style (which increases tone to give bright and vivid colors). Swipe left until you find the style you prefer, and then tap the Use button for that style — for example, tap the Use Vibrant button to use the Vibrant style.

TIP

The Camera app applies your chosen photographic style when you capture each photograph, using the style for processing the photo. You cannot directly remove the style, although you can edit the photo to change its effects. Styles are different from filters, which are applied after processing and which you can remove completely.

- **Prioritize Faster Shooting:** Set this switch on (green) if you will need to shoot photos in rapid succession or capture fast-moving subjects. When the Prioritize Faster Shooting feature is on, the Camera app reduces its processing of the photos you've just taken when needed to allow you to shoot further photos more quickly. When you're shooting quickly, this setting may cost you some image quality, but the difference is usually not noticeable.
- **Lens Correction:** Set this switch on (green) to have the Camera app automatically correct lens distortion on the front camera and on the ultrawide lens on the rear camera.

» **Smart HDR:** Set this switch on (green) to use the Smart HDR feature, which takes multiple exposures and blend their best features into a single photo with high dynamic range.

» **Macro Control:** Set this switch on (green) to have the Camera app automatically display the macro control icon (shown in the margin) for switching to the ultrawide lens to shoot macro (close-up) photos and videos.

When you finish choosing settings in the Settings app, return to the Camera app and test them.

Viewing and Organizing Your Photos

Your iPhone provides the Photos app for viewing, editing, and sharing photos. In this section, we go through viewing and editing, leaving sharing until the final section of the chapter.

Opening the Photos app and navigating to a photo or video

As you've seen earlier in this chapter, you can tap the photo and video viewer in the Camera app to go to the Photos app and make it display the last photo or view you took. If necessary, you can navigate back from that photo or video to other items you took earlier.

Other times, tap the Photos app icon on the Home screen to open the Photos app (see Figure 14-11). You can then navigate to the main areas of the app by tapping the four tabs at the bottom of the screen:

» **Library:** Tap this button to display your entire photo library. You can then navigate by years, months, and days. Tap the Years subtab to display a list of the years your photos cover, and then tap the year you want to view. Within that year, you see a list of the months; tap the month you want. Within that month, you guessed it — you see a list of the days, and can tap the day you want. You then see that day's photos, and can tap a photo to display it full screen.

» **For You:** Tap this button to display the list of memories and featured photos. A *memory* is an automatically curated collection of your photos and videos, typically tied to a date, a person, a location, or a theme. For example, if you frequently go to Wichita and take photos there, the Photos app might create a memory called Visits to Wichita containing its pick of your best photos from there. Typically, a memory includes canned music, transitions between photos, and a visual theme.

» **Albums:** Tap this button to display the Albums screen, which presents your photos neatly divided into albums. The Photos app supports two kinds of albums: regular albums, which contain only the photos you've explicitly added to them; and smart albums, which update themselves automatically based on the criteria defined for them.

The Albums screen contains four or five lists of albums:

- *My Albums:* This list contains two smart albums that Photos maintains: The Recents album contains all your recent photos and videos, and the Favorites album contains all the photos and videos you've marked as favorites. This list also contains each regular album or smart album you've created.
- *Shared Albums:* This list contains albums you're sharing with other people and albums they're sharing with you.
- *People, Pets & Places:* This list contains two smart albums that Photos maintains: The People & Pets album contains photos in which Photos has identified the faces of people or pets, while the Places album contains photos that include geographic location data.

FIGURE 14-11: Tap the Library screen in the Photos app to display your entire photo library.

- *Media Types:* This list contains a smart album for each different media type your library includes: video, selfies, Live Photo, portrait, panoramas, time-lapse, slo-mo, bursts, screenshots, and screen recordings. Photos runs the smart albums. These albums enable you to browse all the media of a particular type easily. For example, if you want to see all the photos you've taken using portrait mode, simply tap the Portrait album.
- *Utilities:* This list contains four smart albums that Photos manages. The Imports album contains each photo or video you've imported. The Duplicates album contains each pair of duplicate photos or videos, enabling you to merge each pair into a single item. The Hidden album contains photos or videos you've hidden to keep them from prying or shockable eyes. And the Recently Deleted album contains each photo or video you've deleted in the last 30 days, giving you plenty of time to recover deletions you've reconsidered.

- **Search:** This screen enables you to search your photos and to browse by People & Pets, by Places, or by categories (such as Vehicles or Performances).

As you can see, the Photos app gives you various ways to reach the photos or videos you want to view. If you want to work with photos or videos you've taken recently, tap the Albums tab, tap the Recents album, browse to the photo or video, and then tap to open it.

From here on, I assume you're working with photos. See the next chapter for information on working with videos.

Viewing a photo

When you open a photo in the Photos app, you'll see controls like those shown on the left in Figure 14-12. These are pretty straightforward:

- **Back:** Tap the < icon to go back to the previous screen, such as the album that contains the photo.
- **Date and time:** This readout shows when the photo was taken.
- **Edit:** Tap this button to open the photo for editing. See the section "Editing Your Photos," later in this chapter.
- **Actions:** Tap the three-dot icon to display the Actions menu, which contains the Copy, Duplicate, Hide, Slideshow, Add to Album, Adjust Date & Time, and Adjust Location commands.
- **Share:** Tap the arrow-escaping-a-box icon to display the Share sheet, which enables you to share the photo via means ranging from AirDrop to Messages, Mail, and Notes. You can tap the Use as Wallpaper button to use the photo as a wallpaper image for your iPhone.
- **Favorite:** Tap the heart icon to mark the photo as a favorite. You can then access it quickly via the Favorites album. Tap again to remove favorite status.
- **Info:** Tap the circled-*i* icon to display the photo's info screen (see the right screen in Figure 14-12). The info includes the camera model, image format (such as HEIF or JPEG), the camera lens used, the resolution and aspect ratio, the ISO, the focal length, the *f* stop, the exposure, and the location. (In the right screen in Figure 14-12, the photo used night mode and a 1.1-second exposure.) You can tap the upper Adjust button to adjust the shooting date and time or tap the lower Adjust button to change the location.
- **Delete:** Tap the trashcan icon to start deleting the photo, and then tap the Delete Photo button in the confirmation dialog.

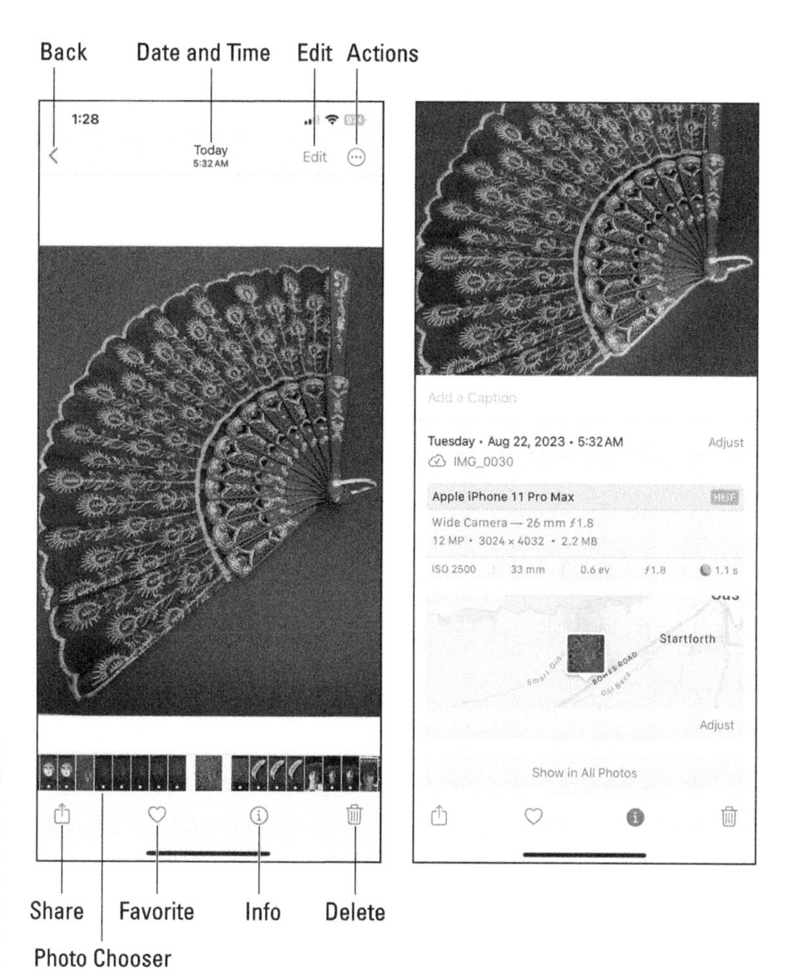

FIGURE 14-12: When viewing a photo (left), you can share it, make it a favorite, delete it, or view its information (right).

Double-tap a photo to zoom in quickly by a standard increment; double-tap again to zoom out. To control the amount of zoom, unpinch with your finger and thumb to zoom in, or pinch with your finger and thumb to zoom out. While zoomed in, drag with two fingers to pan around the photo.

Swipe left on the photo to display the next photo, or swipe right to display the previous photo. To jump to another photo, tap its thumbnail in the Photo Chooser at the bottom of the screen.

Creating an album

To organize your photos, you can create regular albums — that is, albums that don't change unless you change them. Follow these steps:

1. **In the Photos app, tap the Albums tab one or more times until you see the Albums screen, with the My Albums list at the top.**
2. **Tap + (add icon) button in the upper-left corner of the screen, and then tap New Album on the pop-up menu.**

   The New Album dialog opens.
3. **Type a descriptive name for the album, and then tap the Save button.**

   A screen appears for adding photos to the album.
4. **Tap the Photos tab or the Albums tab at the top to specify how you want to navigate to the photos.**

   For example, if you wanted to add photos from the Recents album, you would tap the Albums tab, and then tap the Recents album.
5. **Tap each photo you want to add, placing a blue circle containing a white check mark on it.**
6. **Tap + (add icon) in the upper-right corner of the screen.**

   Photos creates the album and adds those photos to it. The album appears on the My Albums screen, and you can tap it to open it.

After opening the album, you can tap its add icon (+) to start adding other photos to it. You can also open a photo, tap the actions icon (three dots) to display the Actions menu, tap Add to Album, and then tap the destination album.

TIP

As you know, Photos also supports smart albums, which are albums that update themselves automatically based on the criteria you've defined. As of this writing, you can't create smart albums on the iPhone. However, if you have a Mac, you can create smart albums on it, and then sync them to your iPhone via iCloud.

Editing Your Photos

The Photos app provides an impressive range of tools for editing photos:

- **Enhancements:** You can adjust a wide range of settings, such as the exposure, brilliance, highlights, shadows, contrast, brightness, black point, and color saturation.

» **Filters:** You can apply a filter to make a sweeping change to the photo's look and feel.

» **Crop, straighten, and rotate:** You can crop a photo to a different size or aspect ratio, straighten it, or rotate it either horizontally or vertically.

To get started, open the photo you want to edit, and then tap the Edit button. The editing screen appears, normally showing the Adjust tab, which you can see is selected at the bottom of Figure 14-13.

FIGURE 14-13: The Adjust tab enables you to change the color, exposure, brilliance, and other aspects of a photo.

Applying enhancements to a photo

Often, the best way to start improving the colors in a photo is to use the auto tool on the Adjust tab. Tap the auto icon, which shows a wand and three stars, to have Photos apply the changes it calculates are best for the photo. The auto tool adjusts the settings on the enhancements bar as needed. For example, in Figure 14-13, you can see that the auto tool has made a tiny change to the exposure setting and a moderate change to the brilliance setting. There are also other changes you can't see, farther to the right on the enhancements bar.

If the auto tool doesn't have a good effect, tap the Auto button again to remove its changes. You can then work with the buttons on the enhancements bar by scrolling left as needed, tapping the enhancement you want to try, and then adjusting the controls it displays. For example, to change the color saturation, tap the Saturation button, and then drag the slider left to reduce the saturation or right to increase it.

Applying filters to a photo

When you want to make a major change to a photo's looks, you can apply a filter to it. Tap the Filters tab at the bottom of the editing screen, go to the filter chooser below the photo's preview, and then tap the filter you want to apply. You can then drag the slider below the filter chooser left to intensify the filter's effect or right to reduce the effect. Figure 14-14 shows the Vivid Warm filter applied and turned up to maximum.

FIGURE 14-14: Applying a filter on the Filters tab of the editing screen.

Cropping, straightening, and rotating a photo

When you need to crop, straighten, rotate, or flip a photo, tap the Crop tab at the bottom of the editing screen. The controls shown in the left screen in Figure 14-15 appear, and you can perform the following actions:

- **Flip:** Tap this icon to flip the photo horizontally.
- **Rotate:** Tap this icon to rotate the photo counterclockwise.
- **Aspect ratio:** Tap this icon to display controls for adjusting the aspect ratio below the photo's preview. Tap the orientation (portrait or landscape) and the aspect ratio you want.
- **Straighten:** Tap this icon and then drag the slider left or right to adjust the angle of the photo. The left screen in Figure 14-15 gives an example of the changes you can make.
- **Vertical:** Tap this icon, and then drag the slider left or right to rotate the photo's contents vertically.
- **Horizontal:** Tap this icon and then drag the slider left or right to rotate the photo's content horizontally.

» **Crop:** Drag the corners or sides of the crop box to select the area you want to keep, as shown in the right screen in Figure 14-15. You can drag the photo around in the crop box to change the part that cropping will keep.

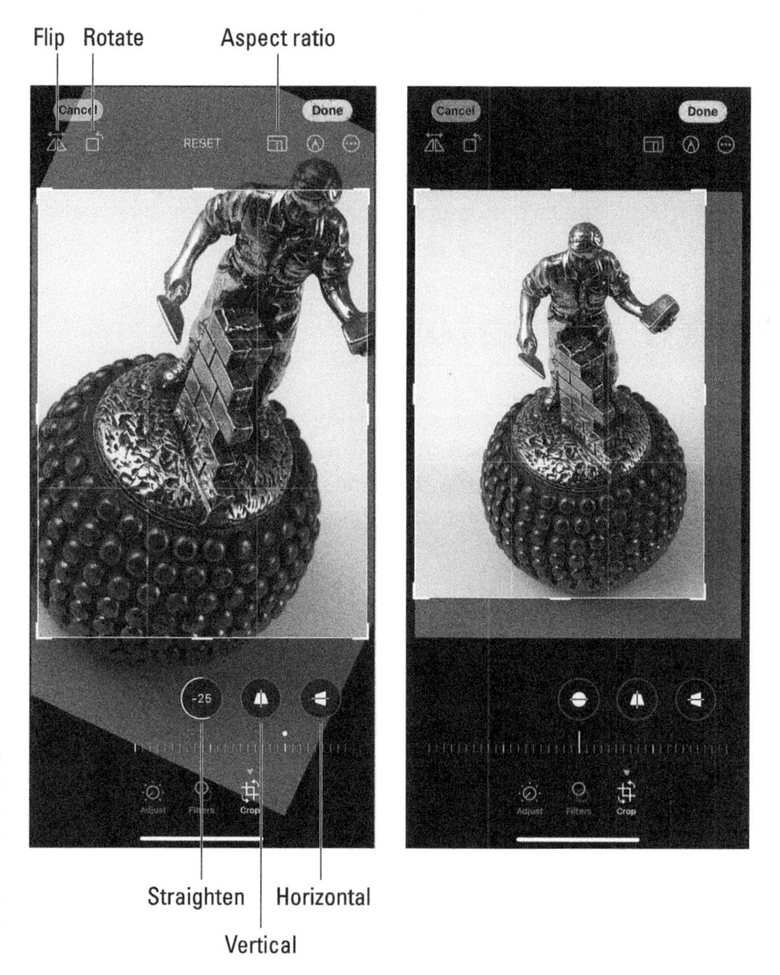

FIGURE 14-15: The Crop tab enables you to rotate or straighten a photo (left) as well as crop it (right).

Copying and pasting your edits to a photo

After editing a photo, you can copy your edits and then paste them onto another photo. This is especially helpful when you make complex edits.

To copy the edits, tap the actions icon (three dots in a circle), and then tap the Copy Edits button on the Options screen that appears. You can then display the next photo that needs the same edits, tap the actions icon, and then tap the Paste Edits button.

Saving your changes to the photo

After editing a photo, save your changes by tapping the Done button.

If you don't want to save your changes, tap the Cancel button, and then tap the Discard Changes button in the confirmation dialog that opens.

Sharing Your Photos

You can share your photos with other people in various ways by tapping the share icon (arrow escaping a box) and working on the Share sheet. For example, you can tap the AirDrop button to share the photo via AirDrop, tap the Messages button to share the photo in an instant message, or tap the Mail button to share the photo in an email message. Earlier chapters explain these methods of sharing, so we won't dig into them here.

What we will explore here are two other means of sharing that are peculiar to Photos: creating a shared album, and creating a shared iCloud photo library.

Creating a shared album

The Photos app enables you to create shared albums, which are albums that share photos and videos with specific people. To create a shared album, follow these steps:

1. **In the Photos app, tap the Albums tab one or more times until you see the Albums screen, with the My Albums list at the top.**
2. **Tap the add icon (+)in the upper-left corner of the screen, and then tap New Shared Album on the pop-up menu.**

   The first iCloud dialog opens, in which you name the shared album.
3. **Type a descriptive name for the shared album, and then tap the Next button to display the second iCloud dialog.**

   In this dialog, you specify the people with whom to share the album.

4. **Add each invitee's address, either by typing it or by tapping the +-in-a-circle and selecting the invitee from your Contacts list.**
5. **Tap the Create button.**

   Photos creates the album and displays the Shared Albums screen.
6. **Tap the new shared album to open it.**

   You see its contents — nothing except an add icon (+).
7. **Tap the add icon (+).**

   A screen appears for adding photos to the album.
8. **Tap the Photos tab or the Albums tab at the top to specify how you want to navigate to the photos.**

   For example, if you wanted to add photos from your Favorites album, you would tap the Albums tab, and then tap the Favorites album.
9. **Tap each photo you want to add.**

   Each photo you tap gets a blue circle containing a white check mark.
10. **Tap the Add button in the upper-right corner of the screen.**

    Another iCloud dialog opens so you can add a message to post with the photos you're adding.
11. **Type a message to accompany the photos.**
12. **Tap the Post button to post the photos and the message.**

    Your subscribers — the invitees who accepted your invitation to the shared album — can now see those photos.

So far, so good — but you should configure a couple of other settings for the shared album. On the screen for the shared album, tap the manage subscribers icon (shown in the margin here) to display the Edit Shared Album screen. Here, set the Subscribers Can Post switch on (green) if you want the subscribers to be able to post photos and comments; if not, set the switch off (white). Further down the screen, set the Notifications switch on (green) or off (white) to control whether or not you receive notifications when subscribers post, like photos, or add comments. Then tap the Done button to save your changes.

Creating your iCloud Shared Photo Library

As well as shared albums, which you can share with anyone online, Photos provides the iCloud Shared Photo Library feature, which enables you to create a shared library in iCloud. This shared library allows sharing with up to five family members or friends. Each member of the group can add, delete, or edit pictures in the library. When they do, iCloud syncs the changes to each member's shared library.

To create your iCloud Shared Photo Library, follow these steps:

1. **Choose Settings ➪ Photos to display the Photos screen in Settings.**
2. **Go to the Library section, and then tap Shared Library Set Up.**

   An information screen appears.
3. **Tap the Get Started button.**

   The Add Participants screen appears.
4. **Tap the Add Participants button, add the participants by typing their names or picking them from your Contacts list, and then tap the Next button.**

   The Move Photos to the Shared Library screen appears.
5. **Specify how to move photos and videos by tapping the All My Photos and Videos button, the Choose by People or Date button, or the Choose Manually button. Tap the Next button, and follow the prompts for choosing.**

   The Preview the Library before Sharing screen appears.
6. **Tap the Preview Shared Library button, make sure the right photos appear, and then tap the Continue button.**
7. **On the Invite to the Shared Library screen, tap the Invite via Messages button or the Share Link button, as appropriate.**

   The Share from Camera screen appears.
8. **Tap the Share Automatically button if you want to share new items automatically via Bluetooth when other members are nearby.**

   For greater control, tap the Share Manually Only button.

   The Your Shared Library Is Ready screen appears.
9. **Tap the Done button.**

IN THIS CHAPTER

» Shooting videos

» Editing footage

» Finding videos

» Playing videos using your iPhone

Chapter 15

Shooting and Watching Videos

As you see in the preceding chapter, the iPhone's diminutive but powerful cameras can shoot video as well as photos. In this chapter, you learn to put your iPhone's video capabilities to use. You also find out where to find videos to play on the iPhone and how to watch them either on the iPhone itself or on a TV to which you connect the iPhone.

Shooting Video

This section shows you how to shoot video using the iPhone's cameras. If you're taking a selfie video, you'll likely want to use the front camera; for just about anything else, the rear camera is clearly better.

Start by opening the Camera app and switching to video mode in one of these ways:

- **Home screen (way 1):** Tap the Camera app icon, and then tap Video on the camera mode bar.
- **Home screen (way 2):** Long-press the Camera app icon, and then tap Video on the pop-up menu.

- **Control Center (way 1):** Tap the Camera icon, and then tap Video on the camera mode bar.
- **Control Center (way 2):** Long-press the Camera icon, and then tap Video on the pop-up menu.
- **Siri:** Say, "Hey Siri, open the Camera app." Then tap Video on the camera mode bar.
- **Lock screen:** Drag the screen from right to left or press the Camera icon. When the Camera app opens, tap Video on the camera mode bar.

Now that you've opened the Camera app and switched to video mode, you need to decide which orientation to use.

Choosing portrait or landscape orientation

You can shoot video in either portrait orientation or landscape orientation. Which orientation to use typically depends on your subject and how you intend to use the video. Portrait orientation works well for some social media sites but will end up letter-boxed (with a thick black bar either side) if you show it on a regular, landscape-orientation TV (unless you rotate the TV. . .). So if you're planning to show the video on a TV, you'll normally want to shoot the video in landscape orientation.

Meeting the video-camera controls

The video-camera controls in the Camera app are largely straightforward (see Figure 15-1). These are the controls, some of which will be familiar from the preceding chapter:

- **Camera active:** This green dot indicates that the camera is active.
- **Camera mode:** On this bar, tap the Video button to switch to video mode.
- **Focus and metering:** The yellow rectangle and light (sun) icon appear briefly when you tap the screen to indicate where to focus and meter the light.
- **Flash:** Tap the lightning bolt icon to toggle the flash between auto and off.
- **Time:** This readout shows the shooting time elapsed.

TIP

- **Resolution:** Tap this button to toggle between 4K and HD resolutions.

  See the section "Choosing settings for shooting video," later in this chapter, for more information on resolution and storage.

- **Frame rate:** Tap this button to cycle through the available frame rates, such as 60, 30, and 24 frames per second (fps).
- **Zoom:** Tap these buttons to switch among the lenses on your iPhone's rear camera. The iPhone in Figure 15-1 has three lenses: 0.5x wide-angle, 1x standard, and 2x telephoto. If your iPhone's rear camera has only a single lens, no buttons appear.
- **Record video:** Tap this button to start recording video. The button changes to the stop recording button, which you tap when you're ready to stop recording.
- **Camera chooser:** Tap this icon to toggle between the main camera and the selfie camera, as usual.
- **Camera and video viewer:** Tap the thumbnail to display the most recent photo or video in the Photos app.
- **Expand:** Tap the arrowhead icon to display the camera controls bar, which enables you to choose other settings, such as exposure and turning action mode on or off.

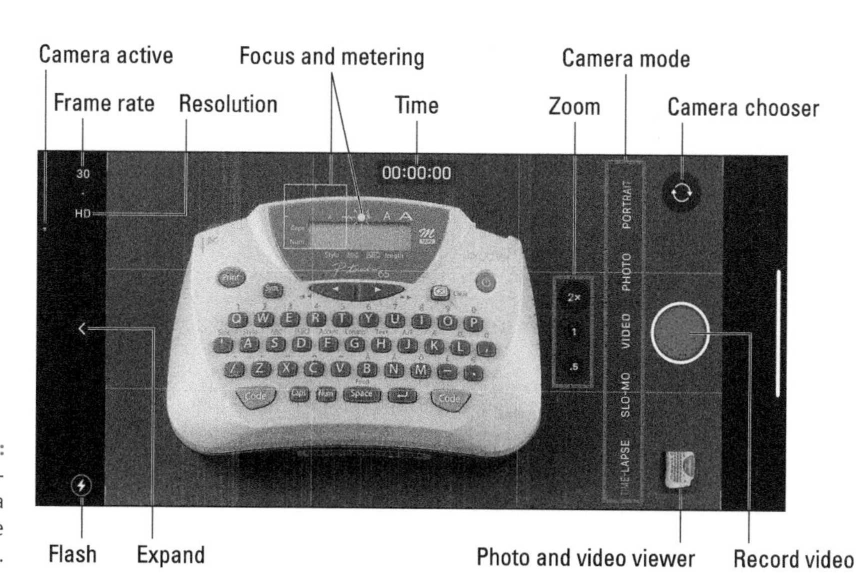

FIGURE 15-1: The video-camera controls in the Camera app.

Taking a video and playing it back

Now that you know what the controls do, try taking a video and playing it back. Follow these steps:

1. **Open the Camera app.**

   For example, tap the Camera icon on the Home screen.

2. **Tap Video on the camera mode bar to switch to video mode.**

   The aspect ratio of the frame changes to the ratio for video capture.

3. **Tap the resolution button to set the resolution, such as 4K or HD.**
4. **Tap the frame rate button to set the frame rate, such as 60 or 30.**
5. **Tap the flash icon if you want to turn the flash on.**
6. **Aim the camera lens at your subject and compose the shot.**
7. **If your iPhone has multiple lenses, tap the lens you want to use, such as 2x.**
8. **Zoom in or out as needed:**

   Zoom out by unpinching or zoom in by pinching. On an iPhone with multiple lenses, you can also zoom by using the zoom wheel: Long-press the current lens button (such as 2x) to display the wheel, and then rotate it to zoom.

9. **Tap the record video button to start recording.**

   The time readout shows the time elapsed.

10. **If you want to take a still photo, take the white take picture button in the lower-right corner of the screen.**
11. **When you're ready to stop recording, tap the stop recording video button, which replaced the record video button.**
12. **Tap the photo and video viewer thumbnail to display the video clip in the Photos app.**

    The clip starts playing automatically (see Figure 15-2). The controls discussed in the following list appear for a few seconds, and then hide if you don't use them, leaving the video playing full screen. Tap the screen when you need to bring the controls back.

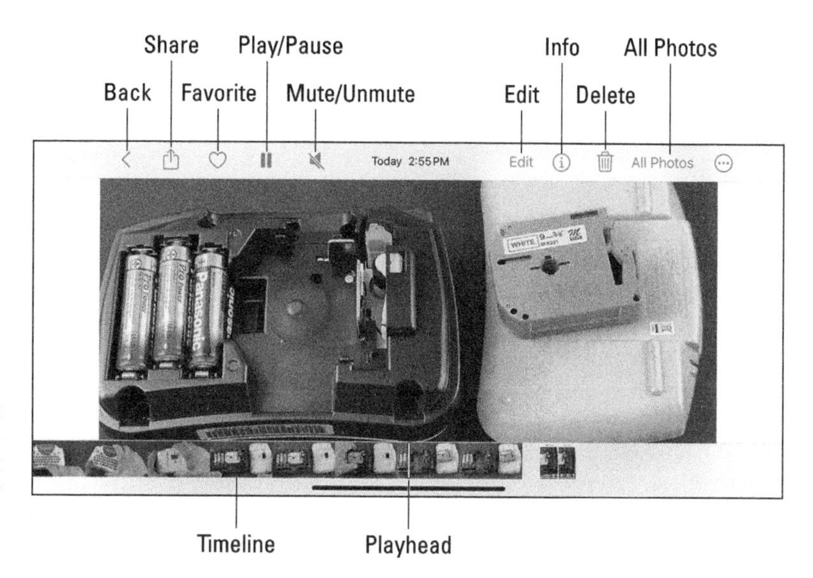

FIGURE 15-2: The video clip plays back automatically in the Photos app.

Here's what you can do with the controls in Figure 15-2:

- **Back:** Tap the < icon to return to the Camera app.
- **Share:** Tap this icon to display the Share sheet, which lets you share the video in various ways, such as via Mail or Messages.
- **Favorite:** Tap this heart icon to mark the video as a favorite. You can then access it easily through the Favorites album.
- **Play/pause:** Tap the play icon to start or resume playback; the icon changes to the pause icon. Tap the pause icon to pause it.
- **Mute/unmute:** Tap this icon to mute or unmute the video's audio.
- **Edit:** Tap this button to edit the video. See the section "Editing your videos," later in this chapter.
- **Info:** Tap the *i*-in-a-circle to display the info screen, which includes the date, time, and location the video was shot; the device used; the lens, focal length, and aperture; and the resolution.
- **Delete:** Tap the trashcan icon to delete the video.
- **All Photos:** Tap this button to display the All Photos album.
- **Timeline:** Tap the timeline to expand it. You can then scroll left or right to move to a different part of the video.
- **Playhead:** This vertical white bar indicates the part of the video that is playing. The playhead remains in place, and you drag the timeline left or right to play a different part of the video.

When you finish viewing your video clip, tap the back icon (<) to return to the Camera app. You can then shoot another clip.

Shooting quickly with QuickTake

The Camera app includes a QuickTake feature designed to enable you to start shooting video without having to switch explicitly to video mode. To use QuickTake, long-press the take photo button instead of tapping it; this changes the take photo button to the record video button. If you're shooting just a short clip, you can simply continue to press the button, and then lift your finger or thumb to stop shooting. Otherwise, drag the button to the lock icon that appears above it (in landscape orientation) or to its right (in portrait orientation), and then lift your finger; tap the stop recording video button when it's time to stop recording.

TIP

QuickTake hardly saves any time over tapping the video button on the camera mode bar and tapping the record video button, but it can be helpful when you're using your iPhone in broad daylight that makes it hard to see the camera mode bar.

Shooting slow-motion video

To shoot slow-motion video, tap Slo-Mo button on the camera mode bar, and then shoot the video as normal. You'll notice that the stop recording video icon (shown in the margin here) is ringed by small white tick marks to indicate that you're using slow motion.

To adjust the slow-motion playback, tap the Photo and Video Viewer thumbnail to display the video in the Photos app, and then tap the Edit button. Drag the vertical bars below the frame viewer (see Figure 15-3) to adjust the slow-motion section as needed, and then tap the Done button.

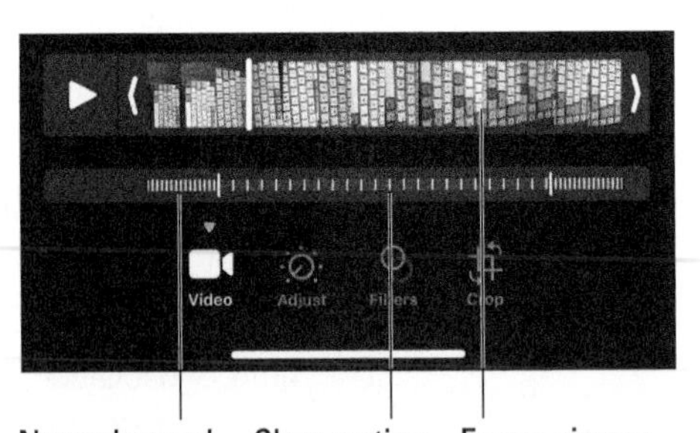

FIGURE 15-3: Drag the vertical bars below the frame viewer to adjust the slow-motion section of a clip.

TIP

The front camera on most current iPhone models can shoot slow-motion video, enabling you to take what Apple calls *slofies.* Tap the Slo-Mo button on the camera mode bar, and then tap the camera chooser icon to switch from the rear camera to the front camera.

Shooting time-lapse videos

To capture a time-lapse photo, tap the time-lapse button on the camera mode bar, and then shoot the video as normal. The stop recording video icon (shown in the margin here) indicates the time lapse by displaying a pattern of tick marks. It draws the pattern on the first iteration, erases it on the second, and then repeats the cycle.

Shooting cinematic video

All current iPhone models except the iPhone SE include cinematic mode, which enables you to add shallow depth of field and focus transitions to the video clips you shoot. To use cinematic mode, tap the Cinematic button on the camera mode bar, and then shoot the video, preferably in good lighting conditions.

As you follow the action, you may notice that if the subject you're focusing on in the foreground turns to look at a person (or even a dog or cat), the focus will automatically shift to that other person (or pet). You can also place the focus manually by tapping where you want it to be. Yellow brackets appear around the subject, the focus sharpens there, and other parts of the frame become blurred.

The Camera app records the focus information used, allowing you to edit the effects afterward. See the section "Editing your videos," later in this chapter.

KEEPING YOUR iPhone STEADY WHILE SHOOTING

The optical image stabilization feature on many iPhone models helps make handheld video easier to watch, but when shooting video, you'll often benefit from mounting your iPhone on a device that will keep it steady. These are your three main choices:

- **Tripod:** When you're shooting from a fixed position, a tripod is hard to beat.
- **Monopod:** When you need to be able to move around but will shoot from a static position, a monopod can be a good solution. Your legs act as the other two legs of the tripod, and you can be ready to shoot the moment you stop moving.
- **Steadicam rig:** A Steadicam is a camera stabilizer mount for shooting video while walking or moving. The camera is mounted on a three-axis gimbal to keep it level, while a counterweight system damps the operator's movements transmitted to the camera.

Choosing settings for shooting video

As you saw earlier in this chapter, you can quickly change the resolution and frame rate directly from the Video screen before shooting a video. But you should also take a minute to configure your default resolution and frame rate, plus other settings for shooting video, in the Settings app.

Choose Settings ⇨ Camera to display the Camera screen in Settings, and then tap the record video button to display the Record Video screen (see Figure 15-4). You can then configure the settings explained in the following list. Exactly which settings are available depends on your iPhone model, and some settings overlap, so you won't see all of these. The settings may also be in a different order:

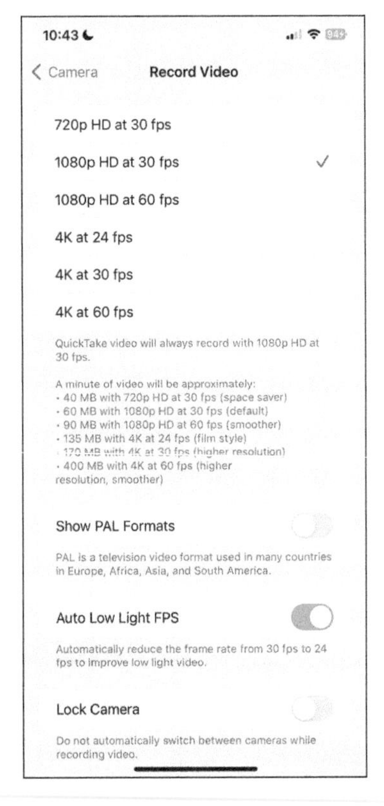

FIGURE 15-4: Configure settings for shooting video on the Record Video screen.

- **Video resolution and frame rate:** In the box at the top, tap the resolution and frame rate you want to use as the default, such as 1080p HD at 30 fps. You can override this setting as needed when shooting video.
- **Show PAL Formats:** Set this switch on (green) if you want the Camera app to display the PAL formats as well as the NTSC formats. PAL stands for Phase Alternating Line and is the television format used widely in Europe, Africa, Asia, and South America. NTSC is the abbreviation for National Television Systems Committee, an analog television broadcasting format used in the U.S., Canada, Mexico, some countries in Central America and South America, and some countries in Asia.
- **Enhanced Stabilization:** Set this switch on (green) to apply enhanced stabilization, which reduces the effects of camera movement by zooming in slightly in video mode and cinematic mode.

- **Action Mode Lower Light:** Set this switch on (green) to use action mode in lower light conditions. This mode reduces video stabilization to take in more light and avoid the video becoming too dark.
- **Auto Low Light FPS:** Set this switch on (green) to allow the Camera app to reduce the frame rate from 30 fps to 24 fps when shooting in low light (again, to avoid the video becoming too dark). Either this setting or the next setting, Auto FPS, appears, not both.
- **Auto FPS:** Tap this button to display the Auto FPS screen, and then tap Off, Auto 30 fps, or Auto 30 & 60 fps to specify when to allow the Camera app to reduce the frame rate either from 30 fps or from both 30 fps and 60 fps to 24 fps when shooting in low light.
- **HDR Video:** Set this switch on (green) to enable video recording in 10-bit High Dynamic Range (HDR) with Dolby Vision. HDR uses multiple exposures to deliver more vibrant color. Using HDR limits the maximum frame rate to 60 fps.
- **Lock Camera:** Set this switch on (green) if you want to prevent the iPhone from switching lenses while shooting a clip. Switching lenses can change the perspective and field of view, disrupt the focus, and cause shifts in exposure, all of which tend to make a clip look less professional. If your iPhone's camera has only one lens, this setting doesn't appear.
- **Lock White Balance:** Set this switch on (green) to lock the white balance value when recording video. Locking the white balance helps keep the video clip's color temperature constant, avoiding potentially distracting color shifts during the clip. That said, it's best to lock the white balance only when the lighting conditions are relatively constant. If you're going to record a clip that starts, say, in bright sunlight and then moves indoors (which would typically have much less light), leaving the white balance unlocked would normally give a better result. Alternatively, shoot the outside clip with the white balance locked, and then shoot the indoors footage as a separate clip, again with the the white balance locked, but this time locked for the different lighting conditions.

Tap the back icon (<) on the Record Video screen to return to the Camera screen, and then tap the Record Slo-Mo button. On the Record Slo-Mo screen, tap the appropriate button, such as the 1080p HD at 120 fps button or the 1080p HD at 240 fps button. As you'd imagine, shooting at 240 fps takes up around twice as much space as shooting at 120 fps. Again, tap the back icon to return to the Camera screen.

Tap the Formats button to display the Formats screen. You visited this screen in the preceding chapter and made a choice in the Camera Capture box at the top. You don't need to change that setting, but see if the Video Capture box appears further down the screen. If so, set the Apple ProRes switch on (green) if you want to shoot video in the ProRes format. ProRes is a popular format for professional video post-production. The disadvantage of shooting ProRes is that video takes up a huge amount of space. One minute of 10-bit HDR ProRes takes about 1.7GB for HD resolution; if you shoot in 4K resolution, one minute takes about 6GB. Once more, tap the back icon to return to the Camera screen once you've finished on the Formats screen.

TIP

The iPhone 15 Pro models can shoot ProRes video directly to an external drive (preferably an SSD) that you connect to the USB-C port using a high-speed USB cable. Exploit this capability if you need to shoot large amounts of ProRes video.

If your iPhone is a 13 Pro or later model, tap the Record Cinematic button to display the Record Cinematic screen. Here, tap the button bearing the resolution and frame rate you want to use for cinematic recording, such as the 1080p HD at 30 fps button, the 4K at 24 fps button, or the 4K at 30 fps button.

Editing your videos

After shooting a video clip, you can edit it directly on your iPhone. The Photos app enables you to trim the video, removing unwanted footage from the beginning and the end; adjust the colors, exposure, and other aspects; apply filters to change the overall look; and crop, straighten, or rotate the video. We'll look at trimming the video here; for coverage of the other editing tools, look to Chapter 14, because these tools work the same way for video as they do for photos.

To open a video clip for editing, first navigate to the clip in the Photos app. If you've just shot the video, you can go straight to it in the Camera app by tapping the photo and video viewer, which shows the thumbnail of the last item you shot. Otherwise, tap the Photos app icon on the Home screen, and then navigate to the video by tapping the Albums tab, followed by either the Recents album (which contains all your recent photos and videos) or the Videos album (which contains all your videos). If you shot the video in slow motion, tap the Slo-Mo album instead.

TIP

Before editing a precious or priceless video clip, duplicate it. With the clip displayed, tap the actions icon (three dots), and then tap Duplicate on the menu that opens.

Once you've displayed the video clip, tap the Edit button to open the clip for editing (see Figure 15-5). Next, tap the Video button to display the trimming controls, and then drag the yellow start point marker (<) and end point marker (>) to select the footage you want to keep. Tap the Done button to display a pop-up menu, and then either tap the Save Video item to save the trimmed clip under its current name or tap the Save Video as New Clip item to save the trimmed clip as a new file, leaving the original untouched.

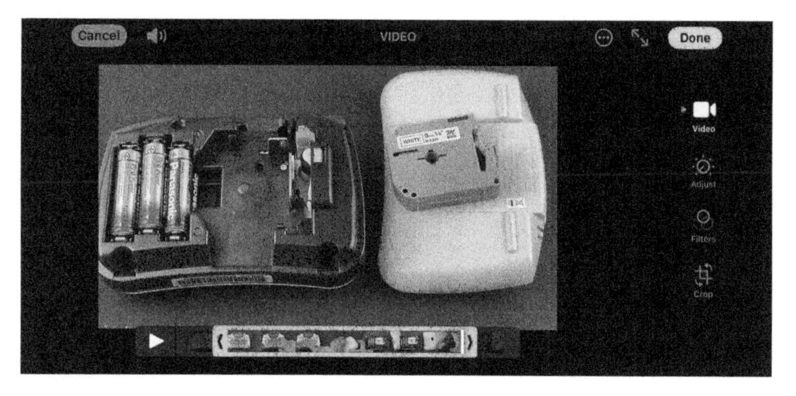

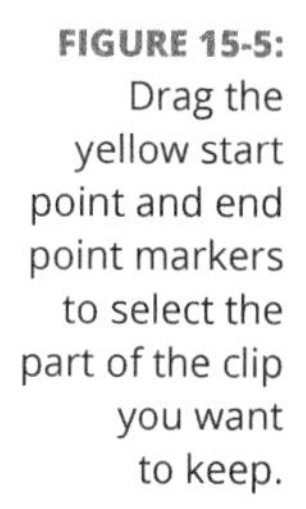

FIGURE 15-5: Drag the yellow start point and end point markers to select the part of the clip you want to keep.

Finding Videos to Watch

Shooting your own videos can be great, and watching them can be even better, especially if you can share the experience with others who enjoy your work. But you'll likely want to watch video content made by others as well. This video content can come from various sources, such as the following:

- **Movies, TV shows, and videos you own:** When you buy a movie, TV show, or other video, you can either download it directly to your iPhone or download it to your computer and sync it to your iPhone.
- **iTunes Store:** Apple's store offers a wide range of TV shows and movies, some for rental and some for purchase.

» **Apple TV+:** Apple's video streaming service offers a range of original content. As of this writing, Apple TV+ costs $6.99 per month after a free seven-day trial, but if you've just bought a new iPhone, iPad, or Mac, you can get three months free. If you subscribe to other Apple services, such as Apple Music or iCloud, look at the various Apple One subscription bundles, which may save you money.

» **Streaming services such as Netflix, HBO Max, Amazon Prime, Disney+, Hulu, Peacock, Paramount+, Showtime Anytime, and YouTube TV:** The apps are free, but the content is tied to a subscription. When you first open the TV app on your iPhone, you can sign in with your cable or TV provider to access content in the apps to which you subscribe.

» **Video podcasts:** Your iPhone's Podcasts app gives you access to assorted video podcasts.

» **YouTube, TikTok, and similar services:** You can either watch these videos on the websites or install the apps on your iPhone. Some services bombard you with ads unless you pay for a subscription.

» **Sports services:** Many sports services are available, via websites, apps, or integration with Apple TV+. For these services, too, you may need a subscription.

Playing Video

Once you've decided what you want to watch, you can watch it easily using the TV app. Launch the TV app by tapping the TV icon on the Home screen, and then tap the Watch Now tab at the bottom of the screen. Tap one of the listings on the Watch Now screen, such as Movies, TV Shows, Sports, or Kids. You see listings of TV shows and movies that you can watch right now, including purchased, rental, and subscription content. You also see other programming that might be of interest listed under Trending, What We're Watching, Free Series Premieres, and so on. Figure 15-6 shows you the Watch Now interface.

You can also access shows and movies you want to watch in your library. You see listings for movies and TV shows, as well as movie posters representing rentals if you've rented any content. If you want to see only the content you've added to this particular iPhone, tap Downloaded. The TV app also displays any recently purchased movies or TV shows in their own sections in the library. Figure 15-7 shows the Library screen.

Here's how to play a video:

FIGURE 15-6:
On the Watch Now screen, explore your viewing options.

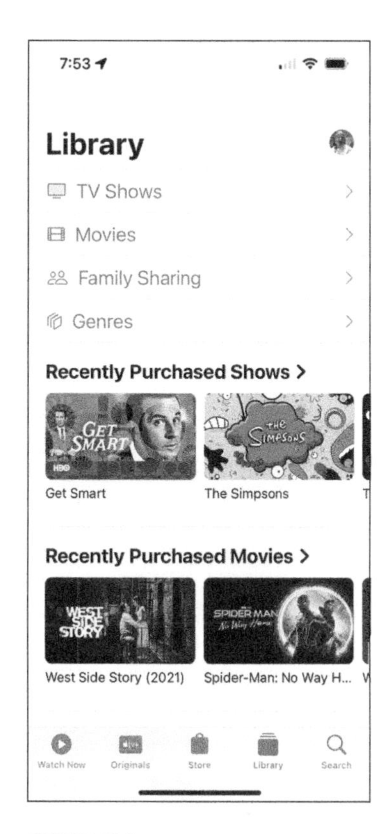

FIGURE 15-7:
Tap Library to get at the content you've purchased or rented.

1. **Tap the Library tab to display the Library screen.**
2. **Tap the Movies item, the TV Shows item, or the Downloaded item, as appropriate.**

   This example uses the Movies item.

   You see poster thumbnails for any movies you previously purchased through iTunes, as shown in Figure 15-8, left. Had you tapped TV Shows instead, you'd get the view in Figure 15-8, right. These posters may appear in alphabetical order by title or sorted by genre.

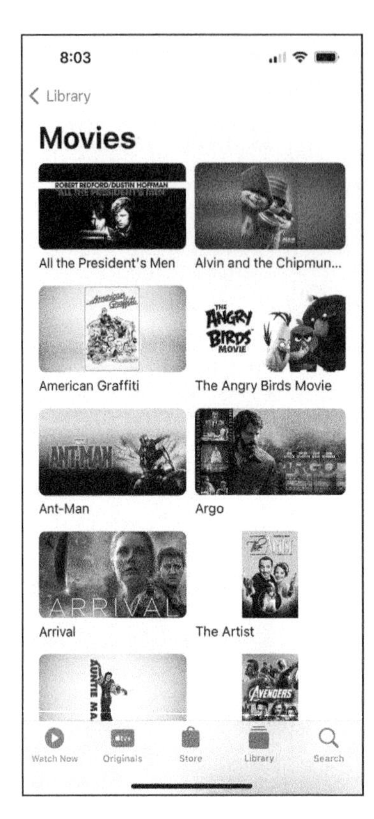

FIGURE 15-8: Choosing the movie (left) or TV show (right) to watch.

3. **Scroll through the list, and then tap the video you want to play.**

   The summary page for the video appears.

4. **Tap the play icon to start the movie playing.**

   If you haven't downloaded the movie or rental you've purchased to your iPhone, the movie starts streaming instead.

5. **Tap the screen to display the controls shown in Figure 15-9.**
6. **Tap the controls that follow, as needed:**
   - Tap the play/pause icon to play or pause the video.
   - Press the volume slider and drag right or left, respectively, to raise or lower the volume. Alternatively, press the physical volume buttons to control the audio levels.
   - Tap the rewind 10 seconds icon (10 in a counterclockwise arrow) to rewind 10 seconds or the fast-forward 10 seconds icon (10 in a clockwise arrow) to move forward 10 seconds.

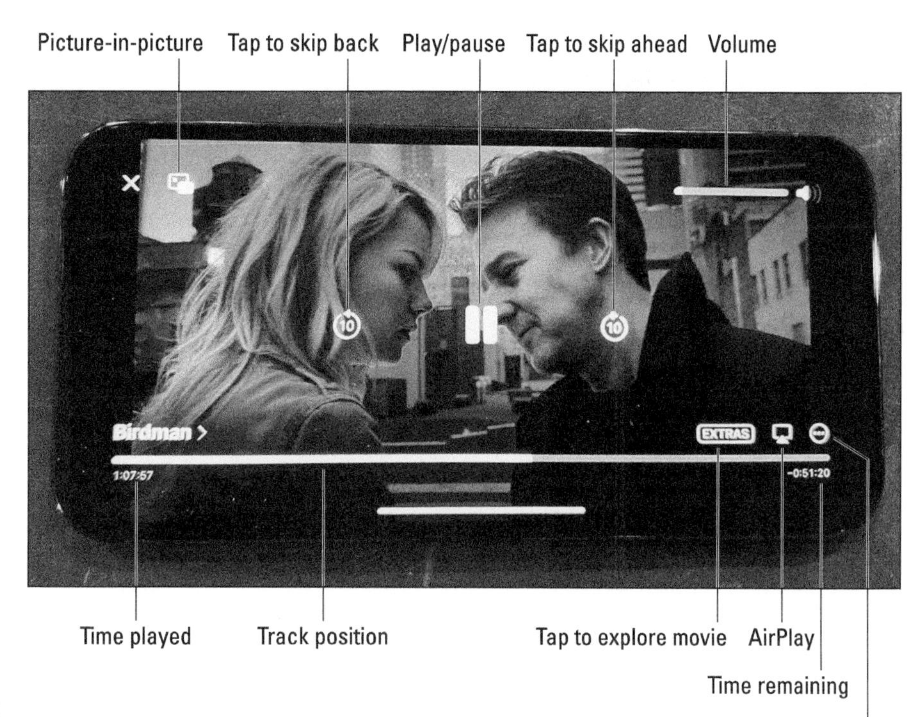

FIGURE 15-9: Controlling the video.

- Press the track position bar and drag right or left to skip ahead or rewind.
- Tap the scale icon to toggle between filling the entire screen with video and fitting the video to the screen. (You won't see this icon in every movie.) Alternatively, double-tap the video to go back and forth between fitting and filling the screen.
- Tap the picture-in-picture icon to continue to watch the video in a smaller window while you're engaged in other apps.
- Tap the three dots icon to summon additional options. You can choose subtitles (if the movie offers subtitles in an available language), change the language of the movie as you are hearing it (if the movie is dubbed in an available language), and change the playback speed (0.5x, 1.25x, 1.5x, or 2.0x, as well as the normal speed).
- Tap the Extras button toward the bottom of the screen to explore specific scenes, bonus material, the cast and crew, related content, and more if such buttons are available for the movie you are watching.

7. **Tap the screen again to make the controls go away (or just wait a few moments until they disappear on their own). Tap again to bring them back.**

8. **Tap the X in the upper-left corner when you've finished watching. (You have to summon the controls back if they're not already present.)**

   You return to the TV's video menu screen.

To remove a downloaded video from the device, tap the Movie poster from the Downloaded listing, tap the word Downloaded from the summary page of the movie, and then tap Remove Download. The movie remains available online.

4 The Part of Tens

IN THIS PART . . .

Learn essential moves for troubleshooting iPhone problems.

Work faster and smarter with ten helpful tips.

IN THIS CHAPTER

» Fixing iPhone issues

» Dealing with network and calling problems

» Sending your iPhone for service

Chapter 16
Ten+ Troubleshooting Moves

Apple makes iPhones and iOS as reliable as possible — but even so, your iPhone may occasionally give you trouble. You might have difficulty making or receiving calls, or the touchscreen might stop responding to your caresses. Connecting to Wi-Fi networks might become problematic, as might syncing with your Mac or PC.

This chapter tells you how to troubleshoot such problems, plus what to do if your iPhone's problems are bad enough that you need to send it in for repairs. Even if your iPhone's working just fine right now, you might want to skim this chapter for future reference.

Update Your iPhone and Computer

Before you try any of the moves in the next section, follow these three steps to bring your iPhone and your computer (if you're using one) up to date:

1. **Install the most recent version of iTunes (for macOS Mojave or earlier, or for Windows) on your Mac or PC. If you're using macOS Catalina or later, make sure you're using the most recent updates to that version.**

   To download the latest version of

   - *iTunes on macOS Mojave or earlier:* Open iTunes and choose iTunes ⇨ Check for Updates.
   - *iTunes on Windows:* Open iTunes and choose Help ⇨ Check for Updates. If Check for Updates doesn't appear on the Help menu, it means you installed iTunes via the Microsoft Store app and must update it via that app: Choose Start ⇨ Microsoft Store to launch the Microsoft Store app, click the Library tab in the left pane, and then click Get Updates.
   - *macOS Catalina, Big Sur, Monterey:* Open System Preferences and click Software Update.
   - *macOS Ventura, Sonoma, or later:* Open System Settings, click General in the sidebar, and then click the Software Update button.

2. **Verify that you're using a USB 2, USB 3, or USB 4 port to connect your iPhone to your computer.**

   On macOS, click , hold down the Option key to make the System Information command replace the About This Mac command on the menu, and then click System Information to open the System Information utility. In the left pane, expand the Hardware list if it's collapsed, and then click the USB item. The USB Device Tree list in the right pane shows the system's USB buses (such as USB 3.0 Bus or USB 3.1 Bus) and the devices connected to them.

   On Windows, click Start, type **device manager**, and then press Enter to open Device Manager. Expand the Universal Serial Buses tree, and then look at the types of USB ports.

3. **Make sure that your iPhone software is up-to-date:**

   - *To check your iPhone software from iTunes on your Mac (Mojave or earlier) or PC:* Connect your iPhone to your computer, launch iTunes (if necessary), and click the iPhone icon near the top left of the iTunes window. Click Summary in the sidebar, and then click the Check for Update button.

- *To check your iPhone software on your Mac running macOS Catalina or later:* Connect your iPhone to your computer, open a Finder window, and select your iPhone in the sidebar. Make sure the General tab near the top is selected, and then click the Check for Updates button.
- *To check your iPhone software from your iPhone:* On the Home screen, tap Settings ⇨ General ⇨ Software Update.

If your iPhone requires an update, follow the updating instructions.

If these three easy steps didn't get you back up and running and your iPhone is still acting up, don't panic and read on.

If your iPhone is hanging, freezing, or otherwise acting weirdly, try the following seven *R*s in sequence, stopping when you've fixed the problem:

- Recharge your iPhone.
- Restart your iPhone.
- Reset (force restart) your iPhone.
- Remove your content.
- Reset your settings and content.
- Restore your iPhone.
- Renew your iPhone with Recovery mode.

The following sections explain these moves in detail.

Recharge Your iPhone

If your iPhone acts up in any way, the first thing you should try is a full battery recharge. Connect it to a power source and give it time to charge.

Make sure to use a power source that delivers enough power, such as these:

- **Mac or PC:** Plug the connector cable into a USB port on the computer itself, not a port on the keyboard or monitor.
- **Powered hub:** A hub that uses an AC power supply is fine. Avoid unpowered hubs.
- **Charger:** Any charger should work well enough in a pinch, but if it's a multi-port charger, use one of the higher-powered ports.

TIP

You can charge your iPhone more quickly by using a higher-powered charger. As long as you use a suitable USB-C-to-USB-C cable (for a USB-port iPhone) or an Apple USB-C-to-Lightning cable (for a Lightning-port iPhone), you can charge your iPhone using Apple's power adapters designed for iPhones, iPads, and MacBooks. See `https://support.apple.com/en-us/HT208137` for more information.

» **Power strip or uninterruptible power supply:** Some of these include powered USB ports for charging your devices.

TIP

If you're in a hurry, charge your iPhone for a minimum of 20 minutes. Longer is better. And for faster charging in any circumstance, turn your iPhone off while it charges.

Restart Your iPhone

If you recharge your iPhone and it still misbehaves, the next thing to try is restarting it.

If the iPhone is responding to the touchscreen, choose Settings ⇨ General, scroll to the very bottom of the screen, and tap Shut Down; then turn your iPhone back on by pressing and holding the Power button. If your iPhone is not responding, restart it the hard(ware) way:

1. **Press the following, depending on your model:**
   - *Face ID:* Press and hold down the side button and either volume button until the slider appears.
   - *Touch ID:* Press and hold down the side button or top button (depending on your model) until the slider appears.
2. **Slide the slider to turn off the iPhone, and then wait a few seconds.**
3. **Turn on your iPhone.**

If your phone is still frozen, misbehaves, or doesn't start up, it's time to try the third *R*, resetting your iPhone.

Reset (Force Restart) Your iPhone

To reset (force restart) your iPhone, follow these steps:

1. **Press and release the volume up button.**
2. **Press and release the volume down button.**
3. **Press and hold the side button until the screen goes dark (around 10 seconds), and then release the side button.**

Forcing your iPhone to restart like this doesn't affect the data it contains, so don't be shy about using this technique. Often, it returns your iPhone to normal.

Remove Your Content

If recharging, restarting, and resetting haven't cured your iPhone's ills, try removing some or all of your data to see if that's what's causing the problem.

WARNING

Before you proceed, back up your iPhone's contents:

- **Mac running Mojave or earlier, or PC:** Click the iPhone icon near the top left of the iTunes window, click Summary in the sidebar, and then click Back Up Now. When the backup finishes, verify that today's date appears in the Backups section.
- **macOS Catalina or later:** Select your iPhone in the sidebar of a Finder window, click the General tab near the top, and then click the Back Up Now button. When the backup is done, make sure today's date appears in the Backups section.
- **iCloud:** Alternatively, initiate a backup to iCloud from your iPhone by tapping Settings ➪ Apple ID ➪ iCloud ➪ iCloud Backup. On the iCloud Backup screen, make sure the Back Up This iPhone switch is set on (green); set the Back Up Over Cellular switch off (white) unless you have an unlimited data plan; and then tap the Back Up Now button.

To remove your iPhone's contents, change your sync preferences so that some or all of your files are removed from the iPhone, and then run the sync. The problem could be contacts, calendar data, songs, photos, videos, or podcasts. If you suspect a particular data type — for example, you suspect your photos because your iPhone freezes whenever you launch the Photos app — try removing that type of data first.

If you don't have a specific suspect, turn off syncing for each type of data, and then run a sync. After that, your iPhone should have next-to-no data on it.

If the problem has now disappeared, restore your data, one type per sync. If the problem returns, experiment to determine which data type or file is causing the problem. If a specific app is causing trouble, delete that app and then reinstall it.

If you're still having problems, the next step is to reset your iPhone's settings.

Reset Your Settings and Content

Resetting involves two steps: The first one, resetting your iPhone settings, resets every iPhone setting to its default — the way the setting was when you took the iPhone out of the box. Resetting the iPhone's settings doesn't erase any of your data or media. The only downside is that you'll have to change some settings back to the way they were afterward.

Just tap the Settings icon on your Home screen, and then tap General ⇨ Transfer or Reset iPhone ⇨ Reset ⇨ Reset All Settings.

If resetting all settings didn't cure your iPhone, try Erase All Content and Settings, but read the next Warning first. Again, choose Settings ⇨ General ⇨ Transfer or Reset iPhone, and then tap Erase All Content and Settings. Because this is such a major step, iOS makes you confirm the erasure twice.

The Erase All Content and Settings command deletes everything from your iPhone — all your data, all your media, and all your settings. To make sure you don't lose anything, sync your iPhone immediately before erasing it.

After using Erase All Content and Settings, set your iPhone up again and see whether it is working properly. If not, restore it, as described next.

Restore Your iPhone

Restoring your iPhone is a drastic step, but it often succeeds after recharging, restarting, resetting, removing content, and resetting settings and content have failed.

To restore your phone, connect it to your computer and back it up. Then:

- **Windows and macOS Mojave and earlier using iTunes:** Click the iPhone icon, click Summary in the iTunes sidebar, and then click the Restore iPhone button.
- **macOS Catalina and later using Finder:** Click your iPhone in the Finder window's sidebar, click the General tab, and then click the Restore iPhone button.

This action erases all your data and media and resets all your iPhone's settings. You can then restore data using iTunes or Finder.

WARNING

After restoring your iPhone to health, you may want to restore its contents from backup. Be careful with this because you might restore whatever caused the problem in the first place.

If restoring your iPhone didn't fix things, try one last thing: Renewing with Recovery mode.

Renew Your iPhone with Recovery Mode

If you've tried all the previous suggestions, or your iPhone is too messed up to try them, try Recovery mode. Follow these steps:

1. **Connect your iPhone to your computer and open iTunes (macOS Mojave or earlier, or Windows) or a Finder window (macOS Catalina or later).**

   TIP

   If you see a battery icon with a red band and an icon displaying a wall plug, an arrow, and a lightning bolt on your iPhone, let it charge for at least 15 minutes. When the battery picture goes away or turns green, proceed to Step 2.

2. **While your iPhone is connected, force restart it with the following steps, but don't release the side button when you see the Apple logo. Instead, wait until the Connect to iTunes screen appears:**

   a. *Press and release the volume up button.*

   b. *Press and release the volume down button.*

   c. *Press and hold down the side button until you see the Connect to iTunes screen, then release the side button.*

3. When you see the option to Restore or Update, choose Update.

Your computer will try to reinstall iOS without erasing your data. Wait while your Mac or PC downloads the software for your iPhone. If the download takes more than 15 minutes and your iPhone exits the Connect to iTunes screen, let the download finish and then repeat Step 2.

4. When the update has finished, sync your iPhone (see Chapter 3) and everything should be well.

WARNING

If you use the Restore option (instead of Update) in Step 3, you may be tempted to use the Restore Backup button to get all your stuff back on your iPhone. That might work, but it may also restore the problem. If it does, Restore again and then sync instead of restoring from backup.

TIP

After you're up and running, immediately make one or two backups of your now-working iPhone — one stored on your Mac or PC and the other stored in iCloud. Just in case.

If your iPhone is still not working at the time, it probably needs to go into the shop for repairs. See the "If You're Still Stuck" section, later in this chapter. And if you erased all your iPhone's content, read the "Dude, Where's My Stuff?" section for suggestions on restoring your files.

Resolving Problems with Calling or Networks

If you're having problems making or receiving calls, sending or receiving SMS text messages, or with Wi-Fi or your wireless carrier's data network, follow these steps:

1. Check the cell signal icon in the status bar at the top of the screen.

If you don't have at least one or two bars, you may not be able to use the phone or messaging function.

2. Open Control Center and make sure airplane mode isn't on.

3. Try moving around to get a better signal.

Moving just a few feet might mean the difference between four bars and zero bars or being able to use a Wi-Fi or wireless data network or not. If you're inside, go outside. If you're outside, move 20 paces in any direction (watch for obstacles). Stop when the iPhone shows the cell signal or Wi-Fi signal has improved.

4. **Try changing your grip on the phone or using a headset.**

   Apple says, "Gripping any mobile phone will result in some attenuation of its antenna performance. . . . This is a fact of life for every wireless phone."

5. **Turn on airplane mode and then turn it off again.**

   Open Control Center and tap the airplane mode icon to turn airplane mode on. If Wi-Fi is still on, tap the icon to turn it off. Wait 15 seconds, and then tap the airplane mode icon again to turn it off again. This move resets both the cellular and Wi-Fi network connections.

6. **Restart your iPhone.**

   If you've forgotten how, refer to the "Restart Your iPhone" section, a few pages back.

7. **If your iPhone uses a physical SIM card, make sure the card is firmly seated.**

If none of the preceding suggestions fixes your network issues, reset your iPhone's network settings. Choose Settings ⇨ General ⇨ Transfer or Reset iPhone ⇨ Reset ⇨ Reset Network Settings. You'll need to set up each Wi-Fi network connection again from scratch afterwards.

Sorting Out Sync, Computer, or iTunes Issues

If you're having trouble syncing your iPhone with your computer, try the following moves:

1. **Recharge your iPhone. See the "Recharge Your iPhone" section, earlier in this chapter.**
2. **Try a different USB port or a different cable.**

   A degraded or damaged USB port or cable can cause sync and connection problems. Check all plugs, connectors, cables, and adapters are inserted fully. Use a USB port on your computer rather than a port on a hub or on a peripheral, such as a keyboard.

3. **Restart your iPhone (see the earlier section "Restart your iPhone"), and sync again.**

4. **Restart your computer. This is a good general troubleshooting move.**
5. **Reinstall iTunes (all but macOS Catalina or later users).**

To make sure iTunes is up to date, choose iTunes ➪ Check for Updates on macOS. On Windows, choose Start ➪ Microsoft Store, click the Library tab in the left pane, and then click Get Updates.

Getting More Help on the Apple Website

Still having problems? Try the Apple website before doing anything rash.

First, turn to the excellent set of support resources at `www.support.apple.com/iphone`. You can browse support issues by category, search for a problem by keyword, or even get personalized help by phone.

Second, go to `https://discussions.apple.com/community/iphone` and check out the discussion communities. They're chock-full of real-world questions and answers from other iPhone users. You can browse by community — Using iPhone, iPhone Hardware, or iPhone Accessories; browse by subcategory; or search by keyword. You'll find thousands of iPhone discussions about almost every aspect of using your iPhone. Better still, frequently you can find the answer to your question or a helpful suggestion.

Third, if you can't find a solution by browsing or searching, post your question in the appropriate Apple community. Check back in a few days (or even in a few hours), and some helpful iPhone user may well have replied with the answer.

Last, try a carefully worded Google (or DuckDuckGo) search. You might just find the solution.

If You're Still Stuck

If you tried all of this chapter's suggestions but your iPhone is still malfunctioning, consider getting Apple to fix it. You can either send in your iPhone or make an appointment online at your nearest Apple Store and go there in person. If your iPhone is still under its one-year limited warranty or you purchased either AppleCare or AppleCare+, the repair will be free — unless you've done something that voids the warranty.

TIP

You can extend your warranty up to two years from the original purchase date by buying the AppleCare+ Protection Plan for your iPhone. You must buy it before your one-year limited warranty expires. The plan costs between $99 and $199, depending upon the iPhone model.

Finally, you can check your warranty coverage on any Apple product at `https://checkcoverage.apple.com/`.

Before you send or take your iPhone in for repair, do the following:

- **Sync your iPhone.** Apple will erase your iPhone during the repair, so sync it with iTunes or Finder first, if you can.
- **Remove any accessories.** Remove the case, screen protector, and your lucky charm.
- **Remove the SIM.** If your iPhone has a physical SIM, pop it out and store it safely.

Although you may be able to send in your iPhone for repair, taking it to your local Apple Store has three advantages:

- **No one knows your iPhone like Apple does.** A Genius may be able to fix whatever is wrong without sending your iPhone elsewhere.
- **The Apple Store might replace your ailing iPhone with a brand-new one on the spot.** This is unlikely but has been known to happen.
- **You can take advantage of Apple's Express Replacement Service (ERS) for iPhones needing repairs.** The AppleCare Express Replacement Service costs $29 when your iPhone is under warranty or covered by an AppleCare Protection Plan. This service provides you with a new or refurbished iPhone when you have to send in your old one for service.

WARNING

If your iPhone *isn't* under warranty or AppleCare, you can still take advantage of the Express Replacement Service, but it costs a lot more. Visit `https://support.apple.com/iphone/repair/service/express-replacement` for current pricing and more information on the ERS service.

If you can't visit an Apple store or a wireless carrier store, call Apple at 1-800-MY-IPHONE (1-800-694-7466) in the United States or visit `www.apple.com/contact` to find the number to call in other countries.

If you choose the AppleCare Express Replacement Service, you don't have to activate the replacement phone, and it has the same phone number as the old phone.

If you have a physical SIM card rather than an eSIM, simply pop the SIM into the new phone, and sync it with iTunes, Finder, or iCloud to fill it with the data and media files that were on your sick iPhone. You'll be good to go.

Dude, Where's My Stuff?

If you performed a restore or had your iPhone replaced or repaired, you still have one more task to accomplish. Your iPhone may work flawlessly at this point, but some or all of your stuff — your music, movies, contacts, iMessages, or whatever — is missing.

Try as many of these three tricks as necessary to get your stuff back:

- **Trick 1: Sync your iPhone, and then sync it again.** That's right — sync and sync again (because sometimes stuff doesn't get synced properly on the first try). If that doesn't set things right, try trick 2.
- **Trick 2: Restore from backup.** If you're using macOS Mojave or earlier or Windows, click the iPhone icon near the top left in the iTunes window, click Summary in the sidebar, and then click Restore Backup. The Restore Backup dialog appears and offers you a choice of backups. Select the one you want, click the Restore button, and let the iPhone work some magic.

  If you're using macOS Catalina or later, connect your iPhone to your computer, open a Finder window, and select your iPhone in the sidebar. Click the General tab near the top of the window, and then click the Restore Backup button.

  TIP

  If you have more than one backup for a device, try the most recent one first. If it doesn't work or you're still missing files, try restoring from each older backup in turn.

  These backups include photos in the Recents folder, text messages, notes, contact favorites, sound settings, and more, but not media such as music, videos, or photos. If media is missing, perform trick 1 again.
- **Trick 3: Sync one media type at a time.** If neither trick 1 nor trick 2 worked, try trick 3. Start by downloading apps you use regularly from the App Store. Test the iPhone for a while and, if everything seems okay, sync another type of media — music, movies, contacts, calendars, photos, or whatever. Test the iPhone again, and if everything seems okay, continue until the problem recurs or you've synced all your media. If the problem returns, you should have a good idea of which media type is to blame.

IN THIS CHAPTER

» Typing faster with QuickPath and hardware keyboards

» Sharing web pages and links

» Creating free ringtones with GarageBand

» Closing apps from the multitasking screen

» Capturing screenshots

Chapter 17
Ten Helpful Hints, Tips, and Shortcuts

Here are ten helpful hints, tips, and shortcuts to help you get the most out of your iPhone — everything from ways to input text faster through closing apps quickly and capturing the contents of the screen.

Use QuickPath for Speed and Accuracy

To enter text faster and more accurately, use the QuickPath keyboard and slide your finger from letter to letter rather than tap each letter in turn. Touch the first letter in the word you want to type and then, *without lifting your finger,* slide your finger to the second letter in the word. Then, *without lifting your finger,* continue sliding to each of the remaining letters in the word.

This feature is called Slide to Type, and it's enabled by default. If it isn't working, choose Settings ➪ General ➪ Keyboard and make sure the Slide to Type switch is set on (green). While you're here, you might also want to set the Delete Slide-to-Type by Word switch on (green) to enable yourself to delete your typed-by-sliding text one word at a time rather than one character at a time.

TIP

As long as the Predictive switch on the Keyboard screen is set on (green), you'll usually see the correct word above the keyboard before you slide onto all its letters. If so, stop sliding and tap the word to enter it.

WARNING

If you start a word with a letter that has alternate forms and you pause on that letter, iOS interprets your action as a long-press and displays the pop-up panel containing those alternate forms (see Figure 17-1). All the vowels and most of the consonants have these alternate forms. You then need to lift your finger and start again.

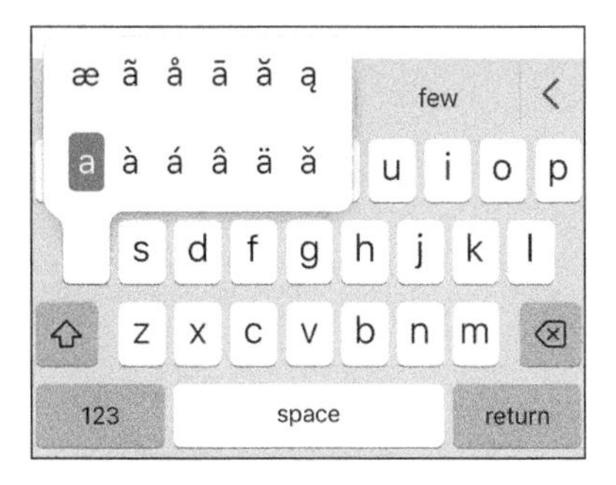

FIGURE 17-1: If your finger lingers on the QuickPath, the panel of alternate characters may pop up.

Autocorrect Is Your Friend

Your iPhone's autocorrection feature can help you type faster and more accurately. Keep these two tips in mind:

- **Let Autocorrect handle contractions.** You can type *dont* to get *don't* or type *cant* to get *can't*. Autocorrect handles most contractions smartly, inserting the apostrophes for you. Autocorrect does pretty well with *its* (possessive, meaning "of it") and *it's* (the contraction for *it is* or *it has*), suggesting the contraction when it seems likely from the context but otherwise leaving the possessive unchanged. Similarly, Autocorrect suggests changing *were* to *we're* if that seems to suit the context, but it struggles with *well* and *we'll*.
- **Reject the Autocorrect suggestion if it's wrong.** This tip applies only if you've set the Predictive switch on the Keyboard screen off (white), which causes Autocorrect suggestions to appear in bubbles instead of on the suggestions bar. If Autocorrect suggests a word you don't want, finish typing the word you do want, and then reject the suggestion by tapping the little *x* on its bubble. Doing this makes your iPhone more likely to accept your word the next time you type it (or less likely to make the same incorrect suggestion the next time you type the word).

Connect a Hardware Keyboard

Your iPhone's on-screen keyboards are as good as Apple can make them, and dictating text into your iPhone via Siri can ratchet up your input speed impressively. But if you need to work seriously with text on your iPhone, there's no substitute for a hardware keyboard.

You have three choices for connecting a hardware keyboard:

- **Bluetooth:** Pretty much any Bluetooth keyboard released in the last 10 years will work with your iPhone. You may occasionally run into a keyboard that uses too old an implementation of Bluetooth; but normally, you can just put the keyboard into pairing mode, choose Settings ➪ Bluetooth on your iPhone, and tap the keyboard's entry on the Bluetooth screen. If the Bluetooth Pairing Request dialog opens, type the pairing code on the keyboard and press the Enter key, and you'll be in business.
- **USB:** Your iPhone can use many USB keyboards (one at a time!) as long as either your iPhone has a USB-C port or you have a Lightning-to-USB connector to provide a USB port on a Lightning-port iPhone. If your iPhone displays a message saying that the keyboard is unsupported because it draws too much power, you're out of luck.
- **Lightning:** Keyboards with Lightning connectors are rare, but if they turn up on Amazon or eBay, you can be confident that they're compatible with your Lightning-port iPhone. If you're a serious keyboard user, a Lightning keyboard can be a great solution.

Once you've connected your keyboard, you can simply start cranking out purple prose as fast as your fingers can flail. But if you need to change the logical layout of the keyboard's keys, choose Settings ➪ General ➪ Keyboard ➪ Hardware Keyboard to display the Hardware Keyboard screen. At the top of the screen, tap the current language to display the language selection screen. You can then choose your layout, such as Dvorak or Colemak.

Also from the Hardware Keyboard screen, you can remap the keyboard's modifier keys by tapping the Modifier Keys button and working on the Modifier Keys screen. So if you want the keyboard's Caps Lock key to act as the globe key for the iPhone, remap it here. You can also set No Action for a modifier key — for example, because you frequently press it by accident.

Happy (real) typing!

Assault on Batteries

Extending battery life — the length of time the phone can run on the battery — is a major concern for most iPhone owners.

WARNING

First: If your iPhone wears a case, charging the iPhone while it's in that case may generate more heat than is healthy. Overheating is bad for both battery capacity and battery life (and bad for your iPhone in general). So if your iPhone feels warmer than usual while charging, you might want to take it out of its case before you charge it. This is easier with some cases than with others.

If you're not using a 3G, 4G, 5G, or Wi-Fi network or a Bluetooth device (such as a headset or car kit), consider turning off the features you don't need in Control Center. Switching the iPhone's functions on and off takes only moments, and doing so could mean the difference between running out of juice and being able to make that important call later in the day.

Better still, enable airplane mode (either in Control Center or in the Settings app) to turn off all your iPhone's radios. Airplane mode disables cellular, Wi-Fi, Bluetooth, Location Services, and push notifications all at once. You can't make or receive phone calls in airplane mode, nor do anything that requires network access. But if you're going to be off the grid for any length of time, like on a long flight, your battery will last a lot longer if you enable airplane mode until you return to civilization.

While you're in Control Center, use its brightness slider to reduce brightness manually whenever you can see the screen well enough at a dimmer setting. You may also want to enable the auto-brightness feature, which lets your iPhone adjust its screen brightness based on current lighting conditions to save battery power. From the Home screen, tap Settings ➪ Accessibility ➪ Display & Text Size, and then tap the Auto-Brightness switch to turn it on. (Some people find auto-brightness too aggressive in turning down the wattage. "Why is my iPhone screen so darn dim?" is a common complaint.)

REMEMBER

You can long-press the Brightness slider in Control Center to summon an overlay containing buttons to enable or disable True Tone, enable or disable night shift, and switch between light and dark mode.

If your iPhone's battery is low and airplane mode and auto-brightness aren't appropriate solutions, try low power mode. Just tap Settings ➪ Battery and set the Low Power Mode switch on (green). With this feature enabled, mail fetch, background app refresh, automatic downloads, and some visual effects are disabled to reduce power consumption.

And while you're on the Battery page in Settings, scroll down and you'll see which apps used the battery in the past 24 hours; you can also tap the Last 10 Days tab to see the last 10 days' usage. On the Last 24 Hours tab, tap a section of the Battery Level histogram to see which apps were active at that time. Down in the Battery Usage by App section, tap Show Activity to see how long each app was active, or tap Show Battery Usage to show how much battery power each app devoured.

Location Services can use lots of battery power, so consider turning this feature off (tap Settings ➪ Privacy & Security ➪ Location Services, and then set the Location Services switch off, so it's white) until you need it. Many apps that use Location Services will offer to enable it for you if you launch those apps while Location Services is turned off.

Believe it or not, disabling EQ (see Chapter 13) for audio also conserves battery life a bit. EQ (equalization) may make music sound a bit better, but it also uses more processing power, and that means it uses more battery. If you've added EQ to tracks in iTunes or the Music app using the Track Info dialog and you want to retain that EQ from iTunes or Music, set the EQ on your iPhone to Flat. Because you're not turning off EQ, your battery life will be slightly worse, but your songs will sound just the way you expect them to sound. Either way, to alter your EQ settings from the Home screen, tap Settings ➪ Music ➪ EQ.

Apple says a properly maintained iPhone battery will retain up to 80 percent of its original capacity after 500 full charge and discharge cycles. You can replace the battery at any time if it no longer holds sufficient charge. Your one-year limited warranty includes the replacement of a defective battery. Coverage jumps to two years with the AppleCare and AppleCare+ Protection Plans. During that one-year limited warranty period or the two-year AppleCare or AppleCare+ warranty period, Apple will replace the battery if it cannot hold a 50% charge or more.

If your iPhone is out of warranty, you can get Apple to replace the battery. The cost varies depending on the model; go to `https://support.apple.com/iphone/repair/battery-replacement` to find out the cost for your iPhone.

Tricks with Links and Phone Numbers

When iOS encounters a phone number (such as 1-123-555-4567), a URL (such as www.example.com), or an email address (such as info@surrealmacs.com) in email or text messages, it formats the text as an action button with blue underlining.

You can tap the action button to make the iPhone take its default action:

- **Phone number:** The iPhone displays a little dialog with two buttons, Call (showing the number) and Cancel. Tap Call to call the number or Cancel to close the dialog.
- **URL:** The iPhone launches your default browser, such as Safari, and displays the web page.
- **Email address:** The iPhone activates Mail and starts a new message to that address.

For more choices, long-press the action button to make the iPhone display a menu:

- **Phone number:** The menu, shown in Figure 17-2, offers buttons for calling the number, sending a message, starting a FaceTime call or a FaceTime audio call, adding the number to Contacts, and simply copying the number.
- **URL:** The menu provides buttons for opening the link, adding it to your Reading List, copying the link, and sharing it.
- **Email address:** The menu has buttons for creating a new mail message, sending an instant message, starting a FaceTime call or a FaceTime Audio call, adding the address to your contacts, and copying the address.

FIGURE 17-2: Long-press an action button to see the full range of actions you can take.

Tap the appropriate button, or tap outside the menu to close it.

TIP

In Safari, you can long-press most images to open a menu appears with four commands: Share, Save to Photos, Copy, and Copy Subject. Tap Share to share this image the usual ways (Messages, Mail, AirDrop, and so on). Tap Save to Photos to save the image to the Photos app. Tap Copy to copy the image to the clipboard so you can paste it into any app that accepts pasted images. Tap Copy Subject to copy the image's subject without its background; you can then paste the subject into any app that accepts images.

Share the Love . . . and the Links

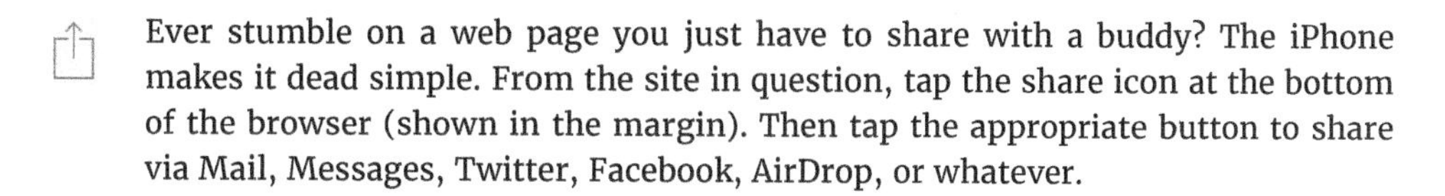

Ever stumble on a web page you just have to share with a buddy? The iPhone makes it dead simple. From the site in question, tap the share icon at the bottom of the browser (shown in the margin). Then tap the appropriate button to share via Mail, Messages, Twitter, Facebook, AirDrop, or whatever.

What happens next depends on which button you tap. For example, if you tap Mail, a new mail message appears with the Subject line prepopulated with the name of the website you're visiting and the body of the message prepopulated with the URL. Just type something in the message body, supply your pal's email address, and then tap the Send button.

And, by the way, the share icon isn't only for web browsers. In addition to finding it in Safari, you'll also find the share icon gracing many apps that create or display content you might want to share.

Don't forget the Shared with You sections in apps including Photos, Safari, Apple News, Music, Podcasts, and TV. They can automatically collect media shared with you by friends. When you can't remember who sent you a link or which app they sent it with, look at the Shared with You section of the appropriate app to find out.

Choose a Home Page for Safari

Unlike most other browsers, Safari on iOS doesn't let you set a home page. Instead, when you launch Safari, it displays the Start page.

To make Safari open a page of your choosing instead, create a *web clip* and add it to the Home screen. Follow these steps:

1. **Open the web page you want to use as your Home page.**
2. **Tap the share icon (shown in the margin) to display the Share sheet.**
3. **Expand the Share sheet by dragging the handle at its top upward.**
4. **Tap the Add to Home Screen button to display the Add to Home Screen overlay.**
5. **(Optional) Change the name of the web clip if you like.**
6. **Tap Add to add the web clip's icon to the first Home screen that has empty space.**

   An icon that will open this page appears on your Home screen (or one of your Home screen pages if you have more than one).

You can now tap this new web clip icon instead of the Safari icon, and Safari opens to your Home page instead of the last page you visited. When you want to open Safari to the Start page, tap the default Safari icon.

If you know that you'll always want to use this icon to launch Safari, replace the default Safari icon on the dock with this icon. Move the default Safari icon to the App Library so you don't tap it by mistake but can easily restore it if you want to use it again.

Create Ringtones for Free in GarageBand

If you have Apple's GarageBand app (a free download in the App Store), you can create custom ringtones. As the source for a ringtone, you can use a sound or song you've created, a song you've ripped from CD, or a song you've downloaded to your iPhone.

Once you've lined up the sound or song you want to use, follow these steps to generate a ringtone:

1. **Launch GarageBand on your iPhone.**
2. **Tap the + in the upper-right corner of the GarageBand Recents screen to start a new GarageBand project.**
3. **Tap the Tracks tab at the top of the window. Swipe left or right until the Audio Recorder project appears, and then tap the Audio Recorder project.**
4. **To use your voice or a recording made with your iPhone's microphone as your ringtone:**
   - **a.** *Tap the OK button.*
   - **b.** *On the dial for microphone sounds, tap the setting you want, such as Clean, Monster, or Telephone.*
   - **c.** *To begin recording, tap the round red record button.*
   - **d.** *When you're finished, tap the square stop button.*
   - **e.** *Skip to Step 6.*
5. **Or to use a song for your ringtone:**

   - **a.** *Tap the timeline icon (shown in the margin) to switch to timeline view.*
   - **b.** *Tap the loop browser icon (shown in Figure 17-3) and then tap the Music tab at the top of the screen to display your Music library.*

c. *Select the song you want to use for your ringtone, press and drag it to the left until the timeline appears, and then release it.*

d. *Tap the region to activate its trim handles, and then drag the handles to set the start and end points for your ringtone, as shown in Figure 17-3. For best results, keep your ringtones under 30 seconds.*

e. *Tap the play icon to hear your work.*

f. *When you're satisfied, tap My Songs to return to the GarageBand Recents screen.*

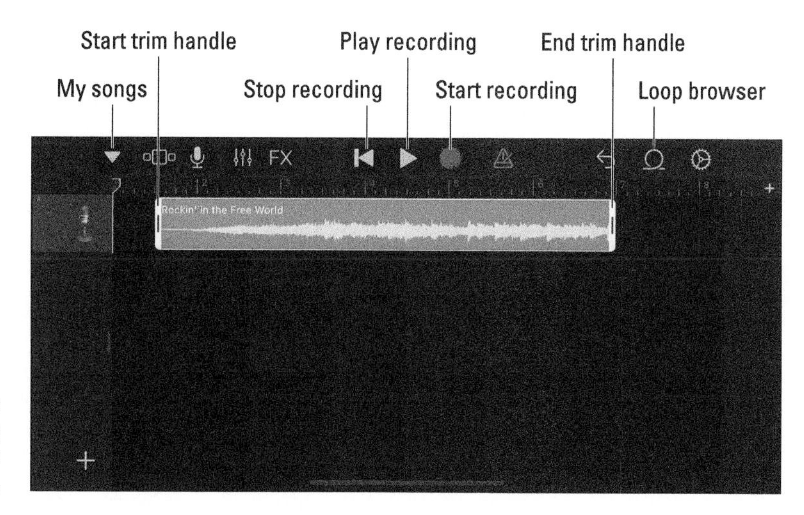

FIGURE 17-3: Creating a custom ringtone.

6. **Long-press the song you just created, and then tap Share in the pop-up menu (scroll down if necessary).**
7. **Tap the Ringtone button to export the song as a ringtone.**
8. **(Optional but useful.) Rename the song.**
9. **Tap Export, in the top-right corner.**
10. **When the Ringtone Export Successful overlay appears, tap Use Sound As to open the New Ringtone dialog.**
11. **Do one of the following:**
    - *To use the ringtone as your iPhone's default ringtone:* Tap Standard Ringtone.
    - *To use the ringtone as your iPhone's default text tone:* Tap Standard Text Tone.
    - *To use the tone for just that contact:* Tap Assign to Contact, select the contact, and then tap Assign as Text Tone or Assign as Ring Tone.
    - *To add the tone to your library without assigning it to a contact or using it as a default tone:* Just tap Done in the Ringtone Export Successful overlay.

Your new ringtone is available on your iPhone. To use it as your ringtone, tap Settings ➪ Sounds & Haptics ➪ Ringtone, and then tap the ringtone in the list of available sounds. To associate the ringtone with a specific contact or contacts, find the contact in either the Contacts app or the Phone app's Contacts tab, tap Ringtone, and then tap the specific ringtone in the list of available ringtones.

You can also assign ringtones as text tones, so you can associate custom ringtones with text messages from a specific contact. The procedure is as just described, but you tap Text Tone instead of Ringtone.

TIP

You don't have to use a song as a ringtone — you can easily use a voice recording instead. Just make sure the recording is available to GarageBand. For example, use the Files app to copy the recording to the GarageBand folder.

Getting Apps Out of App Switcher

Multitasking is great, but sometimes you don't want to see an app's icon in App Switcher. Don't worry — it's easy to remove any app that's cluttering up your screen. Follow these steps to get rid of an app from App Switcher:

1. **Swipe up (Face ID) or double-press the Home button (Touch ID).**

   App Switcher appears. You can swipe the screen from right to left (or left to right) to see additional thumbnails of other open apps.

2. **Slide the thumbnail of the app you want to quit upward and off the top of the screen, as shown in Figure 17-4.**

   iOS closes the app, and it disappears from App Switcher.

3. **Swipe up (Face ID), press the Home button (Touch ID), or tap any app to dismiss App Switcher.**

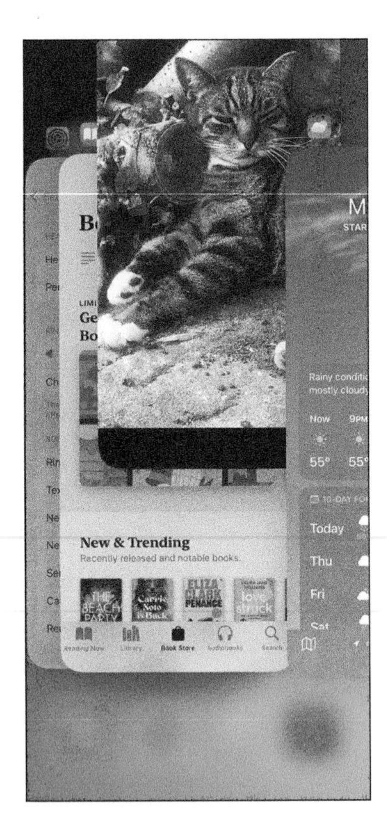

FIGURE 17-4: Slide the app you want to quit (Podcasts in this example) up and off the screen.

TIP

Although you're using this move here to quit an app that you don't want to use for a while, the technique is normally used to forcibly close an app that has stopped responding.

Taking a Snapshot of the Screen

iOS makes it easy to capture what's on the iPhone's screen. This capability is great for quickly capturing everything from fun stuff you run across online to cryptic error messages that flash up on the screen and that you need to dismiss before you can look up what they mean.

To capture the screen:

- **iPhone with Face ID:** Press the side button and the volume up button together.
- **iPhone with Touch ID:** Press the side button and the Home button together.

The iPhone grabs a snapshot of whatever is on the screen, and a thumbnail of the screenshot appears in the lower-left corner of the screen (see Figure 17-5). You can then do three things:

- **Wait.** If you just wait for about 8 seconds, the thumbnail gets bored and exits stage left. At this point, iOS saves the screenshot to the Recents album in the Photo app. From there, you can sync it with your PC, Mac, or iCloud along with all your other pictures.
- **Long-press the thumbnail.** iOS displays the standard Share sheet, enabling you to share the screenshot in its original form.
- **Tap the thumbnail.** iOS displays tools for marking up the screenshot (see Figure 17-6).

FIGURE 17-5:
A thumbnail appears in the lower-left corner of the screen immediately after a screenshot is taken.

FIGURE 17-6:
Tap the thumbnail to mark up the screenshot before you share, save, or delete it.

When you tap the Done button on the markup screen, a menu opens with five choices:

- **Save to Photos:** Tap this button to save the screenshot to your Recents album.
- **Save to Files:** Tap this button to open the Files app so you can browse to where you want to save the screenshot.
- **Save to Quick Note:** Tap this button to start a Quick Note containing this screenshot.
- **Copy and Delete:** Tap this button to have iOS copy the screenshot to the clipboard, so you can paste it elsewhere, and then delete the file. Don't forget to paste the copied screenshot.
- **Delete Screenshot:** Tap this button to delete the screenshot without copying it, passing Go, or collecting $200.

Index

A

B

C

D

E

I

J

K

L

M

N

Q

R

S

T

U

V

W

X

Y

Z

About the Author

Guy Hart-Davis is the author of more than 180 books, including *macOS Sonoma For Dummies, Killer ChatGPT Prompts: Harness the Power of AI for Success and Profit,* and *Teach Yourself VISUALLY iPhone 14.*

Dedication

I dedicate this book to the memory of Robert M. Thomas, a great host, raconteur, and author, whom I was lucky enough to call my friend.

Author's Acknowledgments

My thanks go to the publishing professionals at Wiley who turned a rough manuscript into a finished book. In particular, I thank Steve Hayes for asking me to update this book, Susan Pink for editing the book expertly and making a complex schedule work, Dwight Spivey for performing the technical review with his customary tech chops and good humor, Sofia Malik for keeping a tight grip on the book's production, Ajith Kumar for making the text and figures fit onto pages, and Debbye Butler for proofreading those pages with an eagle eye.

Publisher's Acknowledgments

Executive Editor: Steve Hayes

Project and Copy Editor: Susan Pink

Managing Editors: Sofia Malik and Ajith Kumar

Technical Editor: Dwight Spivey

Proofreader: Debbye Butler

Production Editor: Saikarthick Kumarasamy

Cover Image: © xavierarnau/Getty Images